Maths through play

Easy paths to early learning with your child

by
Rose Griffiths

with foreword by
Carol Baker

Macdonald

A MACDONALD BOOK

© Macdonald & Co (Publishers) Ltd 1988

First published in Great Britain in 1988
by Macdonald & Co (Publishers) Ltd
London & Sydney
A member of Maxwell Pergamon Publishing Corporation plc

All rights reserved.

Printed in Great Britain by Purnell Book Production Ltd
A member of BPCC plc

Macdonald & Co (Publishers) Ltd
Greater London House, Hampstead Rd,
London NW1 7QX

Project Editor	Valerie Bingham
Editor	Susan Baker
Designer	Sally Boothroyd
Illustrator	Sarah Pooley
Picture Researcher	Caroline Smith
Photographer	John Campbell
Production	Rosemary Bishop

Credits
p17 Annie West. *Brinkworth Bear's Opposites Book* (Macdonald)
All photographs are by John Campbell with the following exceptions:
GALT Educational: pp. 56, 57, 58, 81, 89
Sally and Richard Greenhill: p. 47
Barry and Fiona Hunt-Taylor: pp. 44, 91
The publishers and photographer would like to acknowledge the assistance and co-operation of the following schools:
Woodchurch Nursery School
Minster Nursery School
Langtry Nursery School
Franklin D. Roosevelt School.

British Library Cataloguing in Publication Data
Griffiths, Rose
　Maths through play
　　1. Mathematics —— Study and teaching
　　I. Title
　　510'.7　　QA11

ISBN 0-356-13459-8
ISBN 0-356-13460-1 Pbk

Contents

		page
1	Maths is everywhere	6
2	Right from the start	12
3	Counting	22
4	Shapes and spaces	42
5	More number games	64
6	Measuring	80
7	Time	88
	Index	94
	Further reading	96

Foreword

Parents have an important part to play in their children's education. When it comes to reading, for example, we know that parental help makes all the difference to a child's enjoyment and progress. But how can we also help our children become confident and competent in maths?

Unfortunately we are often hampered by our own lack of confidence. We dimly recall incomprehensible formulae or remember all too clearly exercise books covered in red biro. Just how insecure parents – particularly mothers – feel about maths was brought home to me when my daughter's primary school arranged a maths workshop for parents. It was mainly fathers who attended. Mothers had opted out in droves, many telling their partners, 'You go. I'm hopeless at maths.' We cannot be in a good position to show enthusiasm for the subject to our children when we feel so anxious ourselves. And the implications of this role-modelling for girls is particularly worrying since so many develop a negative attitude to maths.

As parents, what we really want are practical ideas for things to do, information about how children learn and above all the reassurance that we are better at maths than we believe. **Maths through Play** meets these needs in a way that is persuasive, down to earth and humorous. Rose Griffiths convinces us that if we can budget, follow a dress or knitting pattern, read a map or estimate the price of a holiday, then we are using mathematical skills.

A child's daily life offers many practical opportunities to learn about number, shape, matching and sorting, etc. For instance, setting places at the table, playing with water, wrapping a parcel are all examples of maths in action. Once we can see the potential in these situations we can give our children a firm practical understanding on which later theoretical learning can be built.

Games enable children to test out and reinforce learning without the risk of boredom. Of the many suggestions in the book for entertaining and easy-to-make games I particularly like Monsters in Dressing Gowns and Secret Sultana Eating.

Children's playing, doing, talking and learning are all interlinked. You can't say where one begins and the other ends. By drawing on *all* these resources we can make maths a relevant and exciting part of our children's lives.

Carol Baker

author of **Reading through Play**

1 Maths is everywhere

I am a maths teacher and a parent too. I know from my own experience that parents can help their children a great deal by laying the foundations of mathematical experiences long before their children start school. Whether you disliked maths at school or loved it, you can help your child by encouraging **maths through play** and in everyday activities at home.

As a teacher, I know that no matter how well I organise the class, or how well I explain things, some children will not get enough individual attention from me, simply because the time available is limited. So, even when your child is in full-time education, encouragement and support from *you* will be of immense value.

Maths is everywhere around us. In this book I hope to point out the mathematics in many of the activities you already do with your child and to suggest ways in which you can extend these. There are plenty of games and activities as well, all tried and tested by mums, dads, childminders and playgroups.

First of all, in case you are thinking of putting this book back on the shelf because the whole idea of mathematics fills you with anxiety, let's take a closer look at a very common feeling amongst parents.

'I'm not at all mathematical . . .'
'. . . so I had better not try to do anything with my child because I'll only confuse matters.'

It is likely that you underestimate your own ability at maths.

You *may* be right to feel you are not mathematical. But perhaps the fault is partly due to the less-than-perfect teaching you had. Many of us had very good maths teachers when we were at school, but not everyone did.

Did you ever feel like this at school?

Perhaps a little more individual attention might have helped you make the most of your mathematical potential.

It is also likely that you use a great deal of mathematics in your everyday life without even realising it.

Can you do any of these?

* Cook a family meal,
* Follow a knitting pattern,
* Read music and play an instrument,
* Play snooker or darts or chess,
* Use a paper pattern to make clothes,
* Follow a map without getting lost too often,
* Do a week's shopping for a family,
* Play whist, rummy or almost any other card game,
* Decorate a room with paint and wallpaper,
* Check a bank statement,
* Use bus or railway timetables,
* Count up the takings from a jumble sale,
* Plan a family holiday.

All of these activities involve mathematical skills. Maths is much more than just arithmetic, and many people find they enjoy some aspects of maths, even if they actively dislike others!

Young children use mathematics in their everyday lives without any prompting. On the next two pages we will look at a typical day in the life of Katie, who is 3 years old.

Katie's Mathematical Day

Is it time to get up? Is it still dark? Is my brother Andy awake yet?
(Time. Making logical deductions.)

Sort out my clothes and put them on the right way round.
(Sorting. Matching. Making decisions.)

Have breakfast and help feed the cat. Not too much! Is that enough?
(Estimating quantity. Comparing.)

Take Andy to school. Go round the corner then cross at the lights.
(Direction. Time.)

Go shopping. How many tins of beans? A large loaf or a small one?
(Counting. Size. Money.)

Put things away. What goes in the fridge? Where do the beans go?
(Sorting. Categorising.)

Watch TV then play with my teddies and jigsaws until it's time for lunch.
(Counting. Size. Shape. Time.)

After lunch we go to the park. Sort out leaves, count ducks, kick my ball, look at shadows.
(Sorting. Counting. Shape.)

Time to collect Andy from school. How far do we have to walk? Are we early or are we on time?
(Time)

Time for tea. What goes on to cook first? Help Andy to lay the table. How many knives do we need?
(Planning. Sorting. Matching. Counting.)

Have a bath and play pouring and splashing. Count all the boats.
(Volume. Counting.)

Bedtime. A story now and two songs. What will we do tomorrow? Don't forget to kiss me goodnight.
(Putting things in order. Making plans.)

Katie's mother or father may never have thought about how her daily experiences are giving her the chance to learn a great deal of mathematics. She is learning in a very informal way in the natural context of everyday situations, and she is enjoying everything she does. Many teachers would love to be able to reproduce these real-life opportunities for learning in the classroom, but this is not usually possible.

You are already providing your child with opportunities to learn many important mathematical ideas, and there are many ways of extending them. The activities suggested in this book do not need any previous mathematical knowledge on your part. In fact, many people find that they only really understand something after they have had to explain it to someone else. It is not at all unusual to find both teachers and parents who say that they have learned more about a particular topic when they were trying to help their children than they ever did as children themselves. So as an added bonus you may find that your own understanding of maths improves as you help your child.

Although each of the chapters which follows looks at a different aspect of maths, you will find that an activity given in any one chapter may also help with other skills and concepts, since there are many connections between one topic and another in maths.

Always keep in mind that children develop at very different rates. You will probably find, though, that the activities in chapters 2, 3, 4 and part of chapters 6 and 7 are most useful with children aged 1 to 5, whilst chapter 5 and part of chapters 6 and 7 are mostly for children aged 4 or 5 and older.

Above all, I hope you will enjoy helping your child discover that maths is fun!

◁ *It can be nerve-racking enough taking just one or two children shopping. Imagine taking a whole class!*

Children learn a great deal about mathematics in everyday situations. ▷

2 Right from the start

Our children are constantly trying to make sense of life. Far from being small, ignorant people who need to go to school before they learn anything, children do their best to try to sort out the puzzling world in front of them from the moment they are born.

In this chapter we will look at some of the ways in which you can help them.

Fortunately, the skills and qualities our children need in order to learn mathematics successfully are also ones which will stand them in good stead in many other aspects of life.

Asking Questions

The art of asking and answering questions sensibly is one which children need time and practice to develop. It is extremely important in maths – not least because some parts of maths rely on understanding one thing properly *before* you move on to the next, and so children who have the confidence to ask questions are less likely to become confused.

Babies enjoy answering questions like 'Where's mummy's nose?'

Let's look for a book about caterpillars in the library.

Babies enjoy pointing out the answer to questions like 'Where's your mummy?', 'Where's your nose?' or 'Where are your eyes?'. Of course, these are all questions to which we adults know the answers already. We are asking them for two reasons: to help the child learn the answers, and for the fun of sharing an activity.

Two year olds can begin to learn to answer more difficult questions. Adults commonly make one of two mistakes at this stage. Sometimes we do not realise that our children have *accidentally* given us the right answers, but do not really understand; or we carry on asking questions which are so easy that they have become boring. It is important to try to reach a balance between providing enough repetition to consolidate your child's learning, and making your questions varied and enjoyable.

Confidence

Once children are two, we start asking them more questions to which we do not already know the answers. These are very important in building up your child's confidence. Quite simple questions will do, as long as you value the replies. 'Would you like one sandwich or two?', 'Where is your teddy?', or 'What shall we do this afternoon?'.

Finding Out

Equally importantly, we should encourage our children to ask questions when they are *not* sure of something. They must learn not to be afraid to say 'I don't know' if they are asked a question they cannot answer. We can do this best by setting a good example. When you say things like 'I can't do this, I'll have to ask Jane what to do' or 'Where's the phone book? I've got to look something up', you are making sure your child knows that it is perfectly normal not to know everything!

Logical Thinking

Mathematics and logical thinking go hand in hand. Sometimes young children do things which seem to be totally illogical, even stupid, but if we could see things from their point of view we would realise that they are actually thinking very hard. **They get things wrong because of their lack of experience of the world** – and in most cases, they learn *because of* their mistakes.

Child's Eye View

Some of us find it easier than others to imagine what is going on in a child's mind. Perhaps you are already an expert at seeing things from a child's eye view. Almost everyone who has children has a favourite story about something comical their child has said or done, as a result of not understanding the adult world. Try to think of something your own child has misunderstood, or perhaps think back to your own childhood for an example – and try to pinpoint the logic behind it.

Here are two 'misunderstandings' from my own family, to start you thinking . . .

Cow Food

When Tanya was four, she asked me when we were going to get our cow. 'Don't be silly,' I said, 'Where would we put a cow? Cows are enormous!' Tanya looked disappointed, but not defeated. 'We must be getting a cow. You've bought some cow food, I've seen it in the cupboard.' It was there indeed – with a beautiful picture of a cow on it, so there was no mistaking who it was for.

'When are we going to get our cow mummy?'

Pink Fish

One of our sons accidentally killed our goldfish by emptying a bottle of baby lotion into their water. He said he did it 'to stop them getting sore and wrinkled.'

Overall, your task is to put yourself in your child's place. Don't jump to conclusions about his or her thinking; instead, ask questions to clarify his or her reasoning, and respond to errors in a positive and helpful way.

Remember that mistakes are a chance to learn something new!

Learning New Words

A great deal of early mathematics relies on learning new words. Let's have a look at how this often happens.

Imagine you are two and you are out in the park with your mother. She says, 'Oh, look! There are the *Pendovers!*' Here is a picture of what you can see:

'Oh look! There are the Pendovers.'

Are the *Pendovers* ✱ the flowers?

✱ the animals?

or ✱ the family?

You really need more clues to decide. Suppose, though, that you decide your mum must be talking about the animals. If you are wrong, you will probably find out when you try to talk about '*pendovers*' to her.

Children under 5 learn new words very fast, but it usually takes a while before the real meaning of a word is clear to them.

When a child learns a new word she or he needs . . .

✱ to hear the word used, and

✱ to use the word

. . . in a variety of ways and in many different contexts. This includes mathematical words as well as other terms.

Many small children over-extend the meaning of a word. For example, they might use 'dog' to mean 'any animal with four legs', or 'square' to mean 'any shape with straight sides'. Sometimes they restrict the meaning too much. Gradually, though, they alter their personal definitions to fit in with the rest of us.

Try to choose your words carefully when you are explaining something to your child, but also be ready to point out that sometimes words have more than one meaning, and that people sometimes use words imprecisely.

Everyday Life

How can you provide enough occasions for your child to talk and think about new words and concepts? **Practical** and **active involvement** in everyday life is one way of doing it. Children need plenty of time to play with household bits and pieces and to help with jobs like laying the table, doing the shopping, feeding pets, making sandwiches, tidying up, and so on.

Any activity which involves carrying out a plan, sorting, matching, comparing, or putting things in order is valuable.

Even if your child is too small to join in with a particular job, you can let her or him watch you and talk about what you are doing.

Just doing the washing will give you dozens of opportunities to talk and listen. For example, you can:

* **Give information**
'The soap powder helps get the clothes clean. We need to get some more.'

* **Encourage discussion**
'How did these trousers get so muddy?'

* **Point out cause and effect**
'Oh, look! Your teeshirt's got a red stain on the front. That's from your lolly – it dripped and made a mark.'

* **Compare and sort things**
'Which socks are the smallest? Which things need ironing and which don't?'

* **Build confidence and independence**
'Can you put these socks in pairs, all by yourself?'

Can you put the socks in pairs all by yourself?

hot   cold

Basic Vocabulary

A good mathematician needs a very broad vocabulary! Start with words like *up, down, bigger, smaller, the same, too much, not enough, above* and *below, in between, opposite, full, empty, long, short, in front, behind, over, under, quickly, slowly* . . . and many others, too. Practise colours when you are getting dressed, or play 'I spy with my little eye something the colour *red*'. (Or try '. . . something which is *stripey*' to practise patterns.)

Books and Photos

There are many good picture books, including board books, which look at important early vocabulary. Any good bookshop or children's library will have a wide selection.

Use your own family photos and books about everyday situations, especially those illustrated with photographs, to encourage your child to think and talk about his or her own experiences.

Talk about pictures in books like these from Brinkworth Bear's Opposites Book.

Toys

Some toys are especially good for developing simple mathematical ideas. *Never* buy something just because the manufacturer recommends it as being educational, though. 'Educational' should never mean 'boring'. The most important questions to ask yourself are whether a toy is *safe* and whether it is *fun* to play with. Here are a few ideas on some favourites for under-threes.

Foam Blocks

These come as beautifully soft cubes in bright colours. Look at them one at a time to see what pictures are on each face. Build them up and knock them down. Sitting and crawling babies love them and toddlers find them easy to build with.

Flip Fingers

Flip the colours in order, name them and count as you flip – one, two, three!

Chime Ball

This ball is a lovely shape and it does not roll away too fast. Turn it round gently. There is a swan, then a horse, then a swan, then a horse, in a pattern.

Hammer and Pegs

Our hammer and pegs set has eight pegs in four different colours. We can sort them into pairs and hammer them down counting 'One, two' over and over again.

Children need time to play on their own with these toys, as well as playing and talking about them with you.

Beakers and Barrels

Sorting ten beakers into order is surprisingly difficult, so less experienced children will do best with only three or four from the set to start with. Show how they stack up, line up and nest together. Use them for tea parties with dolls and teddies. (Do big toys need bigger beakers than little toys?) Play with them in the bath or use them for sand castles. Which sand castle is the biggest and which the smallest?

Barrels are fun to unscrew and screw together, and matching the two halves correctly is hard work at first. Hide small things inside them. Roll them and spin them on the floor. Try to stack them with the smallest barrel at the bottom. What happens when you do this with the beakers?

Make-Believe Play

Sometimes real life does not provide enough time to reflect on important ideas. Make-believe play gives children the chance to re-enact all the experiences which are mathematically useful or interesting in everyday life, in a situation where *they* are in control. In real life, too, children are children – but in their play, they can be mothers, fathers, doctors, bus drivers, shopkeepers or anything else they want to be.

There is no need for costly equipment for make-believe play, but children aged two and over will certainly appreciate a good collection of things to use.

A laundry basket can be used as a doll's bed, a car, a boat, or to keep things in. Upside down, use one as a table.

A teaset is invaluable. If you are going to buy a second teaset, choose a different colour or pattern, so that your child can practise matching things up.

A cupboard or dresser adds variety, as children can then play at putting away shopping and sorting out the teaset and cooking things. Make one by standing a cardboard box on its side, on top of a larger box.

Bits and pieces to furnish the make-believe world could include a toy telephone, a pull-along dog, a set of keys (either real ones or a plastic teething-ring set), and pretend flowers. Small cushions are a good idea, too.

Buy small plastic saucepans to cook with or use the smallest individual sizes of adult pans and baking tins.

A home-made cooker

✶ Find a sturdy cardboard box, (try an off licence!) and seal it closed.

✶ Stick a piece of strong paper on the top of the box, to make a flat surface.

✶ Use a craft knife to cut 3 sides of a door, and a finger hole to open it.

✶ Paint it all over with pale-coloured emulsion, to make it more hard-wearing.

✶ Draw round a small saucer to make 'rings' on the top; paint them with black gloss. Also paint 'controls' on the front.

✶ Make a catch for the door with two split pins and a bit of thick cardboard.

Make a simple cooker and dresser from cardboard boxes.

Tiny washing-bowls and buckets can be bought from camping shops.

Playdough food (made from a recipe which does *not* include a raising agent) is surprisingly strong:

Mix together 2 mugs of plain flour, one mug of salt and a dessertspoon of cooking oil. Add water to make a dough.

Make *small* food shapes (small enough to fit on a doll's teaset plate) and bake them in a very cool oven for an hour or more, until they are hard.

You can make eggs, pasties, samosas, fish fingers, slices of beetroot, sandwiches or almost any favourite food. Do not make anything too thin or small (it will break too easily) nor too big (it may not dry out well in the oven).

Paint them with powder paints or block paints and a coat of polyurethane varnish; or use acrylic paints.

If you want to avoid painting altogether, you could, for example, make a batch of 'carrots' by adding red and yellow food colouring when you mix the dough.

Dolls and teddies are not just for cuddling. They can be talked to, put to bed, and fed, and they never say they don't want to play with you. Dolls' clothes can give practice in matching and comparing sizes, although taking dolls' clothes off seems to be a much more popular activity with the under-fives than putting them on. Choose or make very simple, strong clothes with wide arm, leg and neck holes. Make small bedcovers and pillows, too, or use baby blankets, handkerchiefs and teatowels.

A Pretend Café

Teddies and dolls (and you, parents) can be customers at a pretend café. The café needs a cooker, somewhere to wash up, and plenty of crockery – a teaset and a collection of margarine tubs and yoghurt pots will do. Tables could be cardboard boxes covered with small cloths, or just spread two or three teatowels on the floor, like a picnic.

A till or cash box and some paper napkins will add authenticity. Make plenty of playdough food or cut out thick card to make fried eggs or toast, for example. Long thin yellow stickle-bricks make good bananas! Sort through your child's toys together to get ideas.

Teddies and dolls can be customers at this pretend café. Make an illustrated menu to encourage matching and counting.

A General Store

You can sell anything you want to in a general store. Start by collecting together some small shopping bags and purses, and perhaps something to represent money. With under-fours it may be better to let them hand over 'imaginary' money. Have a box or a toy cash till ready, too, and collect some paper bags.

Collect lots of boxes and cartons small enough to fit in the shopping bags. Very squashy packets can be filled with screwed-up newspaper and sealed with sticky tape – although younger children will often undo them immediately, to see what you have put in there! Try to get at least three or four packets of each kind, to provide enough to make it worth sorting them and counting them. Let your child use some small tins of food from your store cupboard.

You can make tins of food, too. Ask your chemist for some of the grey or white plastic tubs which he or she gets bulk drugs in. Wash them thoroughly, then cut the labels from some real tins of food to fasten round them with sticky tape.

Buy a small selection of plastic fruit (usually sold where you can buy artificial flowers), or make papier mâché fruit and vegetables. Or use the real thing – potatoes and onions survive quite well. Alternatively, sort out some building bricks, pebbles or conkers. Put them in an ice-cream carton or small washing up bowl, and provide a scoop and scales if you can.

Shopping is more fun if you can take it home and put it all away. But take something back to the shop every now and then, to make the play last longer.

The customer should count everything too, to make sure the shopkeeper hasn't made a mistake.

3 Counting

Learning to count properly takes children a surprisingly long time. By providing individual attention and plenty of practice, parents can help at every stage, from when their child first starts to talk, until they are about seven, (or older, for some children).

Counting is very important, as it is the basis for all number work. Children who can count things confidently and accurately will get off to a good start in number work, whereas children who cannot count consistently will be puzzled and confused, especially if they are pushed ahead too quickly by a teacher or parent who has not realised that their counting is weak. Don't give up practising too soon.

Learning to Count

Collect together five or six small objects (for example, toy cars) and ask some children aged 3 or 4 to count them for you, to help you see the problems.

There are four main things for children to learn. Two of them are easy to spot!

Children need to learn the number names 'one, two, three, four . . .'

Kerry is four. She is counting cars. 'One, two, three, four!'

Children need to learn that we count one number for each object, without missing any out or counting any more than once.

What you may not realise is that children also need to learn:

> The answer to the question 'How many?' is just one number; for example, it is 'Three' and not 'One, two, three.'

> The number of objects in a group stays the same, wherever we start counting and however they are arranged.

Learning to count is rather like learning to drive a car or ride a bicycle – it is made up of a group of related skills which develop alongside each other, not necessarily all at the same rate. Sometimes it is helpful to concentrate for a short time on one particular skill, but you do not need to wait until any one skill is perfect before trying to improve another.

Let's look more closely at the four features of counting, and consider some activities to help you and your child concentrate on each one.

The Number Names

Most of us start making our children familiar with the words for numbers when they are tiny babies. We may say, 'One arm, two arms!' when we put on the baby's cardigan, for example, and we sing nursery rhymes with number words in them. If you have ever learnt the counting words in another language, you probably appreciate how difficult it must be for children – a seemingly never-ending string of new words to learn, which people expect you to say in the same order every time. It must be quite a relief to reach twenty one, and to realise that there is a pattern from there on!

At first, concentrate on the numbers one, two, three and four.

Baby's Body

Talk to your baby or toddler while washing, dressing or playing. Count two feet, one mouth, two ears, and so on. Make jokes with older toddlers: 'One leg, two legs, three legs!' (touching an arm when you say 'three legs').

Talk to your baby while dressing him or her. Count arms, legs or toes.

Counting in Rhythm

Count out loud in rhythm while you march about, or bang on a cardboard box or a drum. Count 'One, two, one, two . . .' while you stamp up and down on the spot, or do keep-fit exercises to the count of 'One, two, three, four!'

One, two, three, GO!

Have a race to the next lamp post or to the bottom of the garden. Take it in turns with your child to say 'One, two, three, GO!'

Gradually increase the highest number you count to, as your child gains in confidence.

Scout's Pace

When you want to get somewhere quickly without getting too out-of-breath: walk ten paces then run ten paces then walk ten, and so on. Count out loud along with your child. If you are feeling ambitious, increase the count to fifteen or more each time.

Number Stories and Rhymes

Stories with rhymes can be especially helpful for learning number games. My youngest son's favourites are *'Ten, nine, eight'* by Molly Bang (Picture Puffin) and *'One Bear All Alone'* by Caroline Bucknall (Macmillan).

Read your favourite number stories over and over again.

Fish and Chips

These two numbers rhymes are very effective for teaching number names.

'One, two, three, four, five.
Once I caught a fish alive.
Six, seven, eight, nine, ten.
Then I let it go again.'

Make your hands into fists and bang them together as you say each number in this rhyme:

'One potato, two potato,
Three potato, four.
Five potato, six potato,
Seven potato more.'

You can repeat this rhyme with other words instead of potato. For example, 'One banana, two banana, . . .' or 'One Joanna, two Joanna, . . .'

Counting Everything Once

Young children make more mistakes when they are counting than many adults realise. Even five year olds are remarkably unreliable when counting small collections of only six or seven objects. In any average group of five year olds, less than half of the children are likely to be able to count more than eight objects accurately and consistently.

Young children often make mistakes in *co-ordinating* their counting, giving one number to one object. Older children (and adults) are more likely to make mistakes in deciding which objects they have already counted, and which ones still need counting.

The best way of counting things is to point at them, or to touch them and move them. It is *not* 'more grown up' to just look at what you are counting – yet many children are given this impression, by both parents and teachers. But have *you* ever seen a bank cashier count money without touching it?

Counting Workbooks

The pictures in some counting workbooks are very confusing, and it is much more important to **count real things** than to write in a book. However, if you do use workbooks, tell your child to cross off each object as it is counted, or perhaps to cover each one with a counter, to make sure that everything is counted just once. Choose counting books with very clear illustrations.

Give your child plenty of experience in counting real objects which can be moved as they are counted, and talk about how important it is to **count everything only once.**

Feeding Ducks

Count how many ducks come to you for food, and point out how difficult it is to count the birds because they move about, so you cannot always tell which ones you have already counted.

It would be much easier to count the ducks if we could get them to stand still!

The Answer to the Question 'How Many?'

When you ask a question like 'How many ducks are there?', some children think the answer is 'One, two, three'. They do not yet realise the answer is 'Three'.

When we count a collection of things, we use numbers in two ways. In effect we do this:

In real life, we take a sensible short-cut and instead of using 'first, second, third, . . .' to count with, we use the numbers 'one, two, three, . . .' because it is quicker.

Adults usually count without saying the numbers out loud, and we only say the final number. Children who are very confident at counting will do this, too.

At an earlier stage, you may notice that your child counts 'one, two, three' out loud, but repeats the last number, or perhaps emphasises it. This is a sign that he or she is learning the **'cardinality rule'** – that when we count a collection of things, *the number we finish on* is used to represent the size of that collection.

Help your child by using these three ways to emphasise cardinality:

✳ **Emphasise** the last word in the counting sequence.

One, two, three, **FOUR**.

✳ **Repeat** the last word at least once.

One, two, three, four.
FOUR. There are **four** of them.

✳ **Answer with one number** having counted silently.

How many are there?
Four.

Conservation of Number

The term **conservation of number** is used to describe the fact that *the number of things in a group does not alter* if you move them about or count them in a different order. It seems very obvious to an adult but it is not obvious to a young child.

Children who are not very accurate counters have every reason to believe that the number of objects in a group *can* change – after all, it has happened to groups they have counted! As their counting becomes more accurate, they will agree that the size of a small group does not vary, but they might still be unsure about whether larger groups can change in size.

Every child needs a great deal of independent counting practice to be really convinced of the conservation of number. You can help by using any activity which draws attention to a particular number over and over again.

Egg Box Collection

Use an egg box made for six eggs. Choose six small things to go in the box (the more different they are, the better) and count them carefully along with your child. Take them all out and leave him or her to re-arrange them and count them again and again.

However much you dance, there is still the same number of children.

Children Dancing

This is a good activity for a small group of children who know each other's names. Count how many children are in the group, out loud. Put on some music or sing a song to dance around to, then stop and collect together again. Make sure the children realise that no-one has left and no new person has joined the group by checking everyone by name, then count the group again.

How Many People in Our Family?

Draw all the family or look at a photo or count everyone around the table at a mealtime. How many of us are there? Count again, starting with a different person each time.

Practice Makes Perfect

There are hundreds of ways to encourage your child to count, and the more practice she or he has, the better.

Children need to watch other people counting things, and to count things themselves. We can teach them the number words in the right order, that we make sure we count everything just once, that the last number we count in a group is the answer to the question 'How many?' – and then they can find out for themselves that the number of objects in a group does not change if you rearrange them or count them in a different order.

More ideas follow for games and everyday activities to give both enjoyable and useful reasons for counting.

Sort toy animals into groups and count them.

Animal Counting

Sort toy animals into groups and count them. How many elephants? How many tigers? How many baby animals?

Pass the Potatoes

Count carrots, potatoes or other vegetables when you are getting dinner ready. 'Well, I think we need four big potatoes. Can you pass them to me from the vegetable rack?' Count them again when they are lined up next to the sink.

Laying Tables

There are five people in our family. We need five plates, five forks, five spoons . . . one for each person.

Sultana Snacks

Have a counted snack – three crackers, or fifteen sultanas, for example.

Count on Your Fingers

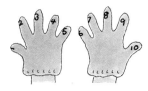

Our number system is based on ten figures, for a very sensible reason! Encourage your child to use his or her fingers as a counting aid. Who wants a cup of tea? How many towels do we need for our swimming trip? Keep a tally on your fingers.

Three Cheers

Celebrate something nice by giving three cheers – keeping check on your fingers if you want to.

'Let's give three cheers for Jenny now she's home from hospital.'

Spot the Dog

How many dogs can you spot when you are out for a walk, on a bus, or in a car?

Playdough Cut-outs

Make pretend dumplings, worms or cut out playdough shapes to count.

This recipe makes a more pliable dough than the one on page 20:

✳ Mix together 2 cups of flour, 1 cup of salt, and 4 heaped teaspoons of cream of tartar, in a large saucepan. (Do not stint on the cream of tartar, or your dough will be too sticky.)

✳ Add 2 tablespoons of cooking oil, 2 cups of water, and food colouring.

✳ Stir over medium heat until the dough forms into one big lump.

✳ Turn the dough out onto a plate to cool.

✳ Soak the saucepan in some soapy water and it will be easy to clean.

✳ Stored in a polythene bag or an airtight box, the dough keeps well.

Going Fishing

A home-made magnetic fishing game is much easier to use than a commercially-produced one, and you can make more fish as your child gets better at counting.

Buy a small horseshoe magnet (some toy shops and most educational suppliers sell them), and tie it onto a piece of string about 45cm long (18 inches). There is no need for a 'rod', particularly not with under fours.

Make fish by cutting shapes out of card. Some children will be able to cut out fish by themselves, and even the tiniest can colour them in (both sides).

Put a paper clip on each fish, and throw them into your 'fish pond'. A washing up bowl is ideal (no water, of course!).

How many fish can you catch in one go? Can you catch four at once? How many fish can you catch altogether?

Count the fish as you throw them back into the pond, as well as when you fish them out.

Rubber Stamps

Use rubber stamps and an ink pad to make counting pictures. Later on, when your child is beginning to recognise written numbers, you could write on a caption for each one, but it is not necessary at first.

✻ Rest your paper on a magazine or newspaper, to make the printing easier.

✻ Draw a simple background for the picture – for example, a pond for frogs.

✻ Count out loud with your child as he or she prints each picture, and then count them again together at the end.

✻ Count the printed pictures again if you want to, as they are coloured in.

Encourage counting in all your family languages.

4 frogs on the pond.

3 rabbits going home.

Family Languages

Does your child speak more than one language? If so, encourage her or him to learn to count in both (or all) your family languages, using any of the counting games and activities in this chapter. If your child is more fluent in one language than another, he or she may find it easiest if you use the most familiar language for the first few times you use any particular activity.

Three of the four main things to learn about counting (see pages 22 and 23) are the same in any language – it is only the number names which are different. Learning to count in more than one language is likely to improve your child's understanding of the counting process.

Every other aspect of maths will also benefit from being discussed in every language your child knows.

Ladybird Counters

Even if you have never considered yourself to be a 'handy' person, you should find these ladybirds and gardens quite easy to make. Children love them, and they are an excellent way of providing counting practice and, later on, of introducing sums. (See Chapter 5.)

To Make the Ladybirds:

You will need

* 40 red bottle tops (from a shop selling wine-making supplies) *or* 40 clean bottle tops and a small tin of red enamel paint,

* A tiny tin of black enamel paint,

* Small paintbrush,

* White spirit (or similar) and jam jar to clean your brush,

* Newspaper.

It is probably not wise to make your ladybirds when there are children around, as enamel paint is difficult to remove!

* Paint all the bottle tops with red enamel, if needed. (Enamel paint gives a very shiny and hardwearing finish.)

* Paint black markings on them.

To Make the Gardens:

You will need

* 4 pieces of thick card, about the size of this book,

* Green felt,

* Fabric glue,

* Black pen, scissors, ruler.

You will also need a box to store the ladybirds and gardens in, and it is worth finding one before you make the gardens, to make sure that you cut your card small enough to fit in. A 2-litre rectangular ice cream tub is ideal.

Thick card is essential, as thin card will curl up. Buy a sheet of red or brown mounting board from an art supplies shop. If you go on a quiet day, they will probably guillotine your four pieces for you. If not, use a sharp craft knife to cut it to size.

* Draw lines on half of each piece of card to represent a brick wall. (If you have a garden with a fence, you could draw fencing instead.)

* Cut a piece of green felt to cover the other half of each card and look like grass. It is easiest to cut the felt a little bit larger than needed: glue it on, leave it to dry, then trim the edges with scissors.

At first, let your child play with the ladybirds in any way he or she wants to. Then try the activities suggested on page 34.

Ladybird Games

The main purpose of having four gardens for the ladybirds is to give your child four chances to practise each task, but they can also be used by two, three or four children playing together.

Start with small numbers (under 5) in each activity, and gradually increase the difficulty. Your aim, of course, is to build confidence – not to catch your child out!

Short sessions of counting practice are best. Five or ten minutes a day is much more help than an hour once a week.

Can you count . . .?

Choose a number of ladybirds to have in each garden. 'Can you count four ladybirds for each garden?' At first it may help if you demonstrate on one garden, then let your child do the other three.

If a group of children are playing together, they can count the ladybirds in each other's gardens to check them.

Tip all the ladybirds back into their box, then try again with another number.

Letting your child throw a dice to decide how many ladybirds to put on the gardens provides more counting practice for numbers one to six.

Ladybird Number Cards

Use self-adhesive red dots (sold in most stationery shops) and small pieces of card to make a set of 11 number cards, for 0, 1, 2, 3, 4, 5, 6, 7, 8, 9 and 10.

Stick on the right number of red dots, then draw on ladybird markings.

To start with, only use the cards for numbers which your child can count accurately. Introduce the others gradually.

Shuffle the cards, or spread them out face down, and pick one at a time to play 'Can you count . . .?'

How Many Ladybirds?

A game for two people. The first person puts some ladybirds on a garden (not more than ten). The second person has to find the right number card to match. Then the second person has a turn at setting out the ladybirds, and the first one has to find the correct number card.

(More ladybird games in chapter 5.)

More Counting Games

You can use these games in the same way as the ladybirds and gardens.

Monsters in the Bath

You will need

* 4 pieces of card, about the size of this book,
* 2 coloured kitchen sponge cloths,
* Pencil, felt pens or crayons, and scissors,
* A box to keep them in.

Draw an outline of a bath with a monster sitting in it, on each of the four cards. Your child could colour them in.

Cut the kitchen sponge into strips about 2cm (¾ inch) wide, then cut each strip into little pieces to make sponges for the monsters to play with in their baths.

Play 'Can you count . . .?' and 'How many *sponges?*' in the same way as the ladybird games (see page 34).

Snakes in the Jungle

You will need

* 4 pieces of card,
* Some smooth thick string and thin string (about a metre of each),
* Felt pens or crayons, and scissors,
* A box to keep them in.

Draw trees, bushes and flowers all over your four cards.

Cut the string into pieces about 4 or 5cm (1½ to 2 inches) long, to make lots of snakes.

Play 'Can you count . . .?' and 'How Many *Snakes?*'.

Perhaps you can think of some ideas of your own, for other counting games!

A Pretend Pet Shop

Visit a real pet shop together to get some ideas first (and to count some animals while you are there, too).

Use soft toys as pets, and make woolly spiders, paper mice and cardboard fish. Find an old tea strainer to be a net for the fish, and a clothes brush to groom furry animals.

Woolly Spiders

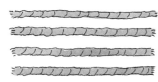

* Cut 4 pieces of thick wool.

* Tie a knot in the middle.

* Count the legs with your child, to make sure there are eight.

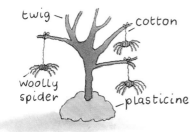

* Hang some spiders from twigs with cotton and sticky tape.

Paper Mice

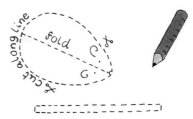

* Cut out a body and tail from paper.

* Draw on eyes, nose and ears.

* Fold body along centre.

* Curl the tail round a pencil first and fasten on to the body with sticky tape.

Collect margarine tubs for customers to take tiny animals home in, and provide cardboard boxes for the rest.

Make tins of dog food and cat food, and packets of other pet food (see page 21). Use playdough (see page 20) to make dog biscuits, and cut out dog bones from thick cardboard. Find some plastic bowls to sell as food bowls or use tin foil dishes.

The shopkeeper can count all the pets regularly to make sure none have got away, and will need to count them when they are sold, too. When there are no customers, the pets will need feeding, fussing and cleaning, and the shop can be tidied up or re-arranged.

Recognising Written Numbers

Let's look more closely at how children begin to recognise written numbers.

Concentrate on 1, 2, and 3 with two and three year olds.

Birthdays

For most children, their own age is the number they are most interested in. Many two year olds can pick out 'my number', and some may recognise the age of a brother or sister. Keep birthday cards with numbers on them, to look at throughout the year.

Look out for your child's age wherever you go. Ignore other numbers at first!

Pat a Cake, Pat a Cake

Pat a cake, pat a cake, baker's man,
Bake me a cake as fast as you can.
Pat it and prick it and mark it with 1,
And put it in the oven until it is done.

Pat a cake, pat a cake, baker's man,
Bake me a cake as fast as you can.
Pat it and prick it and mark it with 2,
And put it in the oven for me and you.

Pat a cake, pat a cake, baker's man,
Bake me a cake as fast as you can.
Pat it and prick it and mark it with 3,
And put it in the oven for baby and me.

Pat your child's hand as you sing this rhyme. Stop when you come to 1, 2 and 3, and draw the number on the palm of your child's hand with your finger.

Number Spotting

You should be able to spot numbers on

* front doors,
* television channel buttons,
* clocks,
* telephones,
* children's clothing,
* calculators,
* calendars,
* car number plates,
* road signs,
* buses,
* lift buttons,

and all sorts of other places.

Their own age is the number that often interests children most.

Number Cut-outs

Draw the number for your child's age on a piece of card about the size of this book, and cut it out. You might need more than one go at this, as it is surprisingly difficult to make a number which looks reasonable! Then let your child colour it, or paint it, or stick paper stars on it to decorate it. Make sure that he or she does not decorate the wrong side of the number, of course.

The decorated number could go on your child's bedroom door, or on a window, or on the bathroom mirror – anywhere it will be noticed. Make cut-out numbers for other children in your family, or even for pets if you want to.

Decorate a cut-out with coloured tissue paper. Tear tissue into small pieces, and screw up each piece into a tiny ball. When you have prepared quite a few, cover the number with glue and stick them on, quite close together.

Crayon Rubbing

This activity works on the same principle as brass rubbing, but you start by making your own 'brass' to rub.

* Cut out a number about 6cm high (2½ inches) from cardboard, glue it onto a rectangle of card, and leave it to dry.

* Fold a piece of paper around the card, and secure it with paperclips.

* Scribble over and over it with wax crayon or chalk. The number will soon show through.

Make several numbers, each on a different card, and use them with your child in two ways.

* Sometimes, let your child look at all the cards to choose a number to rub.

* At other times, wrap all the crayon-rubbing cards in paper before your child sees them, so that they are 'secret'. How quickly can he or she decide what the number is, when scribbling gradually from the top of the number to the bottom?

Make a guessing game of rubbing over numbers.

Telephone Calls
Suggest numbers for your child to dial on a toy telephone, or let him or her help you make a real telephone call. Say each number aloud and point to it.

Playing Cards
Take the picture cards out of a pack of playing cards and put them to one side.

Hunt the Number
Shuffle the remaining cards, then turn the top one over. Suppose it was a six – then your child searches for the other three sixes. Keep going until all the cards are in groups of four.

Cards in Order
Sort out all the cards, ace to five, in one suit, and lay them out in order.

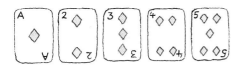

Your child then has to put the ace to five cards from the other suits on top of them.

Repeat with the six to ten cards. If your child finds this easy, try again with ace to ten all in one go, and get your child to help lay out the first suit in order.

Plain Number Cards
Make a set of number cards, like the ones described on page 34 but without the ladybird dots. Use them with the ladybird counters, or with a 'counting collection'.

Counting Collection
Collect buttons, conkers, acorns, seashells, postcards or anything your child likes, to use for counting.

Tea Plate Numbers
Put six tea plates out on the table, and put a number card by each one. Your child has to put the right number of objects on each plate.

Tell your child to shut her or his eyes, and muddle up the numbers. Take it in turns to move objects or numbers until they are all correct again.

Number Towers
Place the number cards in order in a line. Put the correct number of objects in a line next to each card.

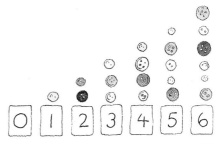

Counting Forwards and Backwards

Some of the activities already described help with practising putting numbers in order. Here are a few more ideas.

Number Frieze

Buy or make a number frieze to put on the wall, and get a simple number jigsaw.

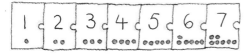

Hop-scotch Trail

Chalk a hop-scotch game or some stepping stones with numbers on the ground, and then hop from one number to another in order, and back again.

Number Train

Stick labels on a toy train, or draw a train on several pieces of card, and practise putting the trucks in order.

Shunt the numbers into order.

Flip Book

Buy a small spiral-bound note book, and write the numbers 0 to 10, one number to a sheet. (If you want to, leave room for your child to draw the corresponding number of objects near each figure.)

Turn the pages one at a time, and ask your child 'What number comes next?' just before you turn over. Try turning backwards as well. 'What number comes before . . .?'

Count Down

Join in with the 'count downs' on some children's television programmes, and sing songs like 'Ten Green Bottles' or 'There were ten in the bed'. You do not have to start with ten, though – five might be better!

Writing Numbers

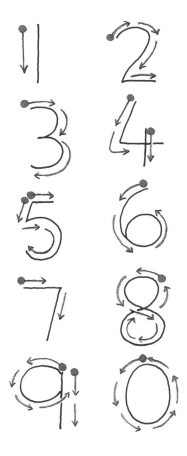

Make sure your child gets into the habit of writing numbers in the correct form.

Some children will want to start writing numbers themselves as soon as they can recognise them. Make sure that they use a conventional method of forming each figure, as it can be very difficult to change habits later on.

Once again, concentrate at first on the child's age. Write the number with your finger, in the flour when you are rolling out pastry, in the sand pit, or in the condensation on a cold window.

Give your child a large felt-tipped pen and a large piece of plain paper, and guide her or his hand when writing the number for the first few times. Then offer encouragement and helpful suggestions as she or he practises.

Reversals

Most children write numbers backwards by mistake at times (especially 2s and 5s). Unfortunately, the more often a child writes a number the wrong way round, the more familiar it begins to seem, and the less it 'looks wrong'. Many problems can be avoided by encouraging your child to look for a number to copy, perhaps on a clock, if he or she is unsure of which way to write it.

I will finish this chapter with both a warning and a reminder.

Remember that children benefit from help at home with counting for much longer than many parents realise – **don't stop practising too soon!**

DO NOT be tempted to start on written 'sums' as soon as your child can write figures. Even teachers feel obliged to do this at times because of pressure from parents; the result for your child may be anxiety and a dislike of number work. **Practical work through play is far more valuable** to your child's education than any amount of confused written work, and a premature start on written work can be harmful.

4 Shapes and spaces

In our everyday lives, most of us use our personal 'feel' for shapes and spaces much more than we use our numerical skills. For example, when you re-arrange the furniture, look something up on a street map, hang a picture or park a car, you use mathematical ideas about shapes and spaces to help you (probably without even realising it). This kind of spatial thinking is also very important when learning mathematics.

In this chapter, we will look at a very wide variety of activities which will help your child develop his or her 'feel' for shape and space.

Many new words and new ideas will arise naturally when your child actively explores the world around you. Even tiny babies love watching things happening around them – and the joy of a newly-walking toddler, who can suddenly see and reach much more of the world, shows us that few children need any prompting to be interested in their surroundings.

Out and About

When you are on a regular journey, whether walking or driving, talk about your route.

* What can you see now?
* What will you see in a moment?
* Which way will you go next?

Remember, though, that a child's eye view is closer to the ground!

'There are all those boxes again. We're nearly at the greengrocer's.'

Whole Body Activities

Climbing, jumping, running, swinging, sliding, rolling and crawling are all enjoyable ways of exploring space. Watch people moving about, dancing or swimming, and talk about what they are doing.

Musical Statues

This is most fun when there are at least 2 or 3 children to join in, as they enjoy watching each other, as well as making statues themselves. You could join in, too.

The children dance about to music, then stop and make a statue whenever you stop the music. Give them special instructions each time. 'Make yourself as tall as you can, like a giraffe'. 'Curl yourself up into a tiny ball', or 'Make yourself as flat as a pancake on the floor'. Admire each 'statue', then start the music again.

Hide and Seek

This is a favourite version for 2 and 3 year olds! Close your eyes while your child hides somewhere in the room. Then pretend you can't find him or her, talking very loudly whilst you 'search'. 'Is she in the cupboard? No! Is she *under* the bed? No! Is she *in* the bed? Yes!'

Chalk Paths

If you have enough space outside, draw a large chalk circle for your child to run or ride around. Try figures of eight and straight and wiggly paths, too. Indoors, use a ball of wool or string to make lines on the floor to follow.

Football

Kick or throw your ball high, low, over your head, through your legs, past the pushchair, not as far as that tree . . . depending on your skill at ball control.

Watch other children moving about and talk about what they are doing.

Jigsaws

To be a good jigsaw-doer, you need to be *observant*, to be able to *match* and *distinguish colours* and *shapes*, to *think logically*, and to be *methodical* and fairly *patient*. All of these attributes are very useful in a budding mathematician, too!

Start with very simple puzzles at quite a young age, and gradually increase the level of difficulty as your child gains experience. Swap puzzles with friends sometimes, to provide variety, but don't give away your easier puzzles too soon – sometimes your child may want to return to them for a while, particularly if he or she is ill or perhaps in need of a boost to his or her confidence. Completing a jigsaw is always rather satisfying.

Me on a Puzzle

Make simple, very personal puzzles for one and two year olds, using photographs or favourite birthday cards.

Cut two pieces of cardboard (the sides of cereal packets will do), exactly the same size as your picture. Glue the picture to one piece of card, to make it more sturdy. Then cut it in two, in a 'V' shape, trying to make the cut in an interesting place on the picture – not necessarily down the middle. Glue one of the two pieces onto the second piece of cardboard, making sure that you match the edges exactly, then leave it to dry. Show your child how the pieces slide together, then let them try.

Make five or six of these puzzles and keep them together in a big envelope.

Inset Puzzles

Most children begin with 4 or 5 piece inset puzzles, where each piece of puzzle is a complete object. Some have pieces graded in size, and these can be very frustrating if the sizes are too similar. Choose puzzles with a clear difference so that it is easier to compare them.

Move on to inset puzzles with more pieces – 10 or 12 pieces, perhaps.

Outline Puzzles

Beginner's outline puzzles usually have only 2, 3 or 4 pieces. Unlike inset puzzles, where you concentrate on the overall shape of each piece, these puzzles need quite a different technique since you have to concentrate on the saw cuts, and sometimes on the colours of the puzzles, to complete them.

Outline puzzles with more than 4 pieces can be surprisingly difficult to do.

Tray Puzzles

Puzzles which fit flush into a frame rarely have pictures with them to refer to, so encourage your child to look at the picture very carefully before she or he tips it out. When choosing these puzzles, check that they have been well cut, as some cheaper ones can be difficult to get in and out of the frame.

Provide a variety of puzzles, gradually increasing the level of difficulty.

Boxed Puzzles

Children of 3 and over who have practised with easier jigsaws will enjoy 20, 25 and 30 piece boxed jigsaws. Some 4 year olds can do a 50 piece puzzle, especially if they are given adult encouragement.

Look for pictures with a lot of clues – plenty of different figures, for example. Try some puzzles where the picture on the box is the same size as the puzzle, and some where it is smaller.

Once your child is able to tackle puzzles with more than 30 pieces, you will soon reach the point where he or she cannot always complete a puzzle in one sitting, either through lack of time or because of flagging concentration. If you need to clear away, it is worth buying a small piece of hardboard about 60cm × 60cm. (24″ × 24″) to work on; it will slide under a settee or a bed until it is needed next. The hardboard will also be useful sometimes as a flat base for building bricks.

Ways of Helping

The best way of helping your child with any task will depend on your child's mood that day and on the type of activity it is. Here are three tactics to try.

Help Stage by Stage
Break the task down into smaller stages and give help at the start of each one. Set 'targets' which are not too far away. For example, 'Can you find two more pieces of jigsaw which fit, before we have our dinner?' rather than 'Can you finish it completely?'

Working Backwards
Often, the beginning of a task is the hardest part. For example, the first few pieces of a jigsaw are usually much harder to do than the last few. Help your child by nearly finishing the task, and let him or her do the last part. Next time, you stop work a bit earlier, so that the child does more.

Help and Encouragement
We all appreciate being told how well we are doing when we are working; children are no exception to this. Point out how much progress your child has already made, and be encouraging. Offer to help if needed – but be careful not to take over.

Jigsaws can be very frustrating so offer your child a little help and a lot of encouragement.

Touching and Looking

Encourage your child to touch and hold things wherever possible to find out more about their shape and texture, as well as looking at them carefully.

What's in the Bag?
Take turns in hiding something in a shoe bag or a pillow case, then trying to decide what it is by touch alone.

Sand and Water
Make hand prints and footprints in sand, or with wet hands and feet on a dry path. Are the prints of your two hands or feet the same? (Younger children will not be able to remember left and right, but they can usually see that there *is* a difference.) Can you put each hand or foot back in *exactly* the same place, on top of your prints?

Let your child explore the differences between wet and dry sand.

Shapes and Spaces
Fill buckets or yoghurt pots with sand and turn them out. What happens with dry sand? What happens with wet sand? Look at and feel the insides of the pots, and compare them with your sandcastles. Try fitting the pots back over the sandcastles, too.

Look carefully at jellies and the jelly moulds they were made in. Make ice cubes and look at the ice trays and the ice. Freeze water in small (freezer-proof) dishes or pots, or make small containers from foil to use, and compare the ice with its moulds.

When you cut out shapes from pastry, look at the holes you leave, and try to fit the shapes back in again.

Patterns

Playing with beads, threading reels, and pegs and pegboards can help children to **recognise**, **invent** and **copy** patterns.

Use large **wooden or plastic beads** with several long and short laces. If possible, store them all in a wide but fairly shallow container, and leave one bead tied on to the end of each lace when you pack away, ready for your child to play with next time.

Always stay close by when he or she is playing with beads, as even children who are normally very sensible seem sometimes to be taken over by urges to put them in ears, noses or mouths!

Macaroni and some pasta shapes can be threaded by older children, using fairly stiff thread or a piece of fishing line. Don't try to use a needle – macaroni is not usually straight enough!

Threading reels are good for building pyramids and towers, as well as for threading. It is useful to have some very long laces with them, such as those made for football and baseball boots.

Make 'snakes' with repeating patterns and see if your child can copy them. Let your child make a snake, then you copy it, and he or she can make sure you have done it properly. Make crowns and necklaces, too.

Pegboards can be used to make patterns in more than one direction. A board with big pegs where the pegs also stack on top of each other, is very satisfying to use, particularly for under fours.

Buy a bag of extra pegs if you can, so that there are plenty to choose from, and put them on a tray so that they do not roll away while your child is playing.

Beads, buttons, cotton reels and pasta can be threaded in repeating patterns.

Try patterns like these on your pegboard.

Sewing

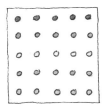
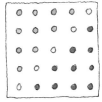

Sewing gives children practical experience of *front* and *back*, *up*, *down* and *through*, and of following a line.

Commercially-produced sewing cards are often much too difficult for young children but it is easy to make your own.

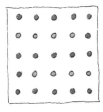
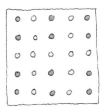

✻ Draw a simple shape onto a small piece of card (about postcard size is usually best). Choose a shape which stays close to the edge of the card. For example, an apple is better than a butterfly because it is more difficult to sew near the centre of the card.

✻ Make holes, using a thin nail or a tapestry needle, about every 1cm (½ in) around your shape.

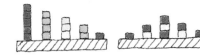

✻ Give your child brightly coloured wool and a tapestry needle (size 20 is good) to sew round.

✻ Fasten the wool on and off with sticky tape on the back of the card.

✻ Your child can colour in the picture with crayons or pencils to finish it off.

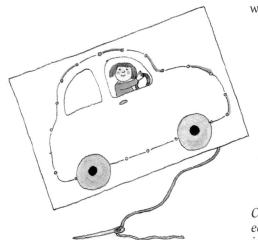

Choose a shape which stays close to the edge of the card. For example, the apple is better than the butterfly.

49

Shape Names

Everything we can see has a shape – but some objects have special names to describe their shape. Which shape names should you introduce to your child? Let's look at 2-dimensional ones first.

Begin with circles.

Circles

Play games where everyone goes round in a circle, like 'Ring a Ring o' Roses' and 'Here We Go Round the Mulberry Bush'.

Look for circles at home. You can see circles on dinner plates, records, clocks, buttons, rings, and cans of food. Run your finger round the top of a mug in a complete circle. Look for very tiny circles (the top of a pin) and very big ones (the rings around a gasometer).

The people on this ferris wheel go round in a circle.

Wheels are circular. How many different things with wheels can you see when you are out for a walk? Cars, pushchairs, wheelchairs, buses, trolleys . . .

Triangles

A triangle is a shape with 3 straight sides. Children who can count to three usually enjoy looking for triangles.

Triangles are strong. Fasten a triangle together from Brio (page 57) or something similar, and make a 4-sided shape as well. Now try to push each of them out of shape.

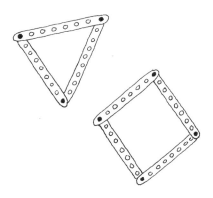

You can squash this four-sided shape, but not the triangle.

Triangles are very important in building because they stay rigid. Look at scaffolding, electricity pylons, and many bicycle and pushchair frames for examples of useful triangles.

Cut a rectangle in half to make a triangle. Do this with sandwiches or toast, and cut sheets of paper in half diagonally to make triangular sheets to draw or paint pictures on.

Rectangles
A rectangle is a shape with 4 straight sides and 4 right-angled corners.

A square is a special kind of rectangle, with all 4 sides exactly the same length.

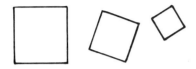

These are all squares.

This shape is not a rectangle since its corners are not right angles.

Many of the shapes we see in everyday life are rectangles.

The ten holes in this shape-sorting ball include a circle, triangle, oval, star, square, pentagon (5 sides), and a hexagon (6 sides).

Shape-sorting Toys
At first, a toy with holes for just three or four shapes (and with more than one of each shape) is best.

The more advanced shape-sorting ball shown above is a very versatile toy. Babies enjoy rolling it around; it rattles nicely if you put a few of the shapes in it. You can help younger children by telling them whether the hole they want is on the red half or the blue half of the ball.

The shapes can be stacked on top of each other and used with building bricks. They are also good cutters for playdough or pastry, *and* we have used them with thick paint for printing.

Most children need adult help to open the ball and shake the shapes out again.

Building Bricks

A varied collection of building bricks will include many geometric shapes which have special names. On the whole, though, you will probably find you and your child use the names **cube** and **cylinder**, and call everything else a brick.

Some older children may be interested to know more names.

Cubes have 6 faces all of which are squares.

Cylinders can be short or long, fat or thin.

These are **cuboids**, or rectangular bricks.

These are **triangular prisms**, or triangular bricks.

◁ Children can build a world of their own from bricks.

A very small child's favourite way of playing with bricks is often just to knock down other people's towers. As your child grows, though, she or he will gradually become better co-ordinated, better able to build interesting structures, and better at coping with the inevitable frustration when things fall down accidentally.

Because building with bricks relies on balancing them (unlike Lego or other fit-together bricks), it helps to have a reasonably large space to play in so that children can crawl around without knocking things over.

Playing on the floor gives you an aerial view of what you have built. Encourage your child to build on the table sometimes, where his or her view will be closer to eye-level.

Provide other things to play with, along with the bricks. For example, plastic dinosaurs need brick mountains to climb and caves to live in. Supply different things on different occasions. Animals, people, doll's house furniture and cars all give children ideas for building.

Talk to your child about his or her building; join in if you want to, make suggestions, and copy each other's ideas. One of the good things about using bricks and other construction toys is that you can try something out and alter it as many times as you want to – unlike drawing or painting, where your creative efforts are more difficult to change.

Discussion can help children concentrate, and encourage them to continue working for longer.

Construction Toys

Construction toys have traditionally been a favourite present for boys, but they are such a useful source of mathematical experiences that both parents and teachers should make sure that the girls get a chance to play with them, too.

The number of good building toys available increases every year and many are certainly much more versatile than the toys *we* had as children. One result of this is that many adults, particularly women, have never played with things like Lego, Brio and Mobilo, and so we feel unsure about what to do with them. Sometimes this means our children cannot make the best possible use of these toys, so let's start by looking at four basic 'golden rules' for construction toys.

Quantity is important. It is difficult to be imaginative and creative with a small amount of equipment. Imagine what it would be like if someone asked you to cook a meal but only gave you three or four ingredients!

Quality is important, too. Good construction toys invariably seem to be very expensive, but many children will use them for at least 6 or 7 years, so they are actually very good value. Some pieces will inevitably get broken, chewed or lost as time passes, and those of us who buy second-hand bags of equipment will probably have started with some rather dubious-looking pieces. *Throw away* any broken pieces and supplement your set with extra wheels or other spare parts whenever you can afford to. Be wary of buying cheap imitations if you can't test them first. Badly fitting pieces can spoil a whole structure.

Storage is a sadly neglected area in many households! Your construction toys should be easy to get out and easy to put away, so that children are encouraged to play with them, even for a short time. Don't insist that structures are broken down into components too soon.

The original box is rarely a good permanent container, although it may be worth cutting out any pictures of finished models from it to keep for reference. Ideally, your storage container should be quite a bit bigger than the equipment seems to need, for two reasons. Firstly, it makes it easier for your child to rummage about, looking for a particular piece. Secondly, it may encourage you to add to the set if you have already got plenty of room to keep things in!

Shallow containers are best. We keep our Lego loose in the bottom drawers of two chests of drawers – Duplo Lego in one drawer, little Lego in the other – and we get the drawers completely out when needed. Big washing up bowls, wide-based jute shopping bags, and large plastic stacking storage boxes are other suitable containers.

Company is a very useful ingredient for good building – someone (adult or child) to talk to about what you are doing. Construction is usually a process of planning what you would like to do and solving problems along the way, so talking can be a great help.

> Ideally, every child should have the opportunity to play with several different sorts of construction toys for fairly long periods of time. Some of these may be at playgroup, nursery or school, and some at home.

If you have two or three different sorts of construction toys it is tempting to tell your child never to get more than one kind out at a time, because the thought of clearing them up at the end is rather daunting. It can be very valuable, though, to compare things you can make with each one. Different sets often complement each other, as each is good for a different purpose. Putting them away again will give your child extra practice at sorting, anyway!

These excellent construction toys are available in most toy shops.

There is usually more than one way to do something.

New Tasks

Let your child discover as much as possible about a particular construction set without interference from you, as it is very satisfying to invent your own methods of doing things, and that will help your child be more confident about tackling new tasks in the future. It is also important to avoid giving your child the impression that there is always a 'right way' to do something.

Which Toys?

I have chosen just four good construction toys to mention here, out of the many useful and interesting ones available. Each of them contributes something different to a child's mathematical experience. They can all be used to make simple and more complex models, so they suit beginners and more experienced builders.

Duplo and Lego

Duplo includes bricks, base plates, vehicles, animals, people and a railway. It is strong, attractive and easy to use.

Lego can be used to make buildings, cars, animals, monsters and people. Every model involves finding the right sizes and shapes of bricks to make it, and working out the best way of fixing them together. As children become more confident builders they can start to think about using colour and symmetry to create patterns in their models.

Later on (from about age 5, with your help), your child can gain valuable experience in following diagrams to build particular models, and in adapting them when needed.

Lego

Duplo

Linking Cubes

'Flexilink Cubes' and 'Multilink Cubes' are both sets of interlocking plastic 'cubes' in ten colours, which are often used to help with number work. They link together on all six faces, so you can use them to make monsters, animals, and so on, all with moveable parts.

Brio-Mec

Brio's construction set is often popular at first because of the hammer, pliers and screwdriver included, though most children seem to abandon the tools and just use their fingers, once they are absorbed in making things. The wooden pieces do get rather grubby. (Do *not* try to wash them, though – sanding them with fine sandpaper is more effective.)

Brio's plastic connectors allow parts to move, and children can learn how to use this movement to make their models more interesting, and they can learn how to prevent movement when it is not wanted (see page 50).

The set's green wheels are interesting. Try making a car, but fix the wheels on through one of the outside holes. What happens?

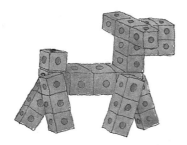

Add extra bricks to its neck and legs to make this dog into a giraffe.

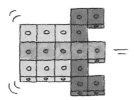

Make a robot which turns into a space ship when you fold its arms and legs up.

This model demonstrates why we fasten wheels on through their centres.

57

Roads and Railways

Road-making and railway tracks provide an interesting way of looking at straight and curved lines and the spaces around them. A complete circuit for your cars or trains is unnecessarily restricting, especially if you only have a small amount of track. It is also not at all like real life – they usually go from place to place, not round in a loop! Build mountains, stations, houses and complete towns on each route, particularly at the beginning and end of the track.

Mobilo

Mobilo is made from smooth and durable plastic. It is bright and attractive, easy to clean, and strong. It has very good connectors which allow you to turn or bend parts of your construction, and so it provides an excellent basis for future work on angles.

The main parts are made from frameworks of triangles and squares, and you can see through the models once you have made them, which is interesting.

Make fire engines with turntables and folding ladders, or birds with flapping wings. Can you make a monster with arms and legs which fold up, so that it looks like a rock when it is folded?

Here is the only layout you can make using 8 curves and 2 straight pieces if you want a complete circuit. But try layouts like these as well.

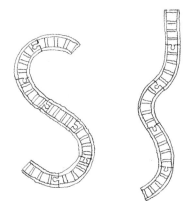

You can find repeating patterns on wallpaper and some wrapping paper.

Patterns Everywhere

Look for patterns wherever you go. At home, in shops and in the street you can find simple patterns in floor tiles and paving, brickwork, railings and fences. Wallpapers, curtain fabrics and some carpets and wrapping papers have repeating patterns. Choose a distinctive part of a pattern and ask your child to find the places where it is repeated.

Look for reflections, too, in mirrors, in water and on other shiny surfaces. Stand models you have made in front of a mirror, so that you can look at the front and the back without turning them round. With over-fives, try to make a model (from Lego, for example) which looks exactly the same from the front as it does from the back.

Natural patterns are important, too. Look at tomatoes, apples and onions when you have cut them in half from top to bottom and compare them when cut across from side to side. Examine any growing things, big or small. What is the same about each plant or animal, and what is different?

Printing

Make up your own patterns or pictures with prints from vegetables or other small objects.

Try vegetables first. Cut small potatoes in half, and use a sharp knife to cut potatoes into other shapes, too.

(One of my children says this is not potato printing, it is chip printing!)

Carrots are good for circles, semi-circles and quarter-circles in various sizes.

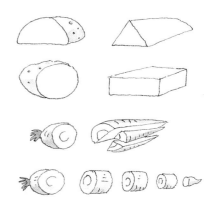

Cut shapes for printing from vegetables.

Mix some powder paint in a jar (make it quite thick) or use ready-mixed paint. Try with just one colour at first, to emphasise the different shapes you are using, and introduce 2 or 3 other colours later on.

Vary the colours of paper to print on. Absorbent paper, such as sugar paper, is best with younger children. Rest your child's paper on a wad of newspaper to make printing easier.

Use a paintbrush to paint over each shape as you use it, then press it down firmly onto the paper. (No banging! That usually causes smudges.) This works much better than dipping the shape into the paint, as the paint is spread more evenly. It also has the advantage that your child has to look at the shape for much longer, as she or he puts paint on.

Try lots of other things to print with. Look at each other's prints and try to decide which things made which prints.

What did we use to make each print?

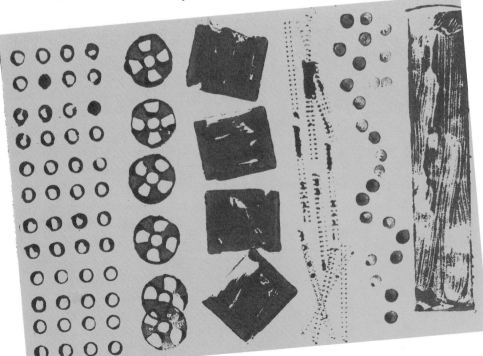

Printing Blocks

Make your own simple printing blocks. Cut shapes out of thin kitchen sponge, cork tile, carpet underlay or felt, and glue each one onto a small offcut of wood or a thick piece of cardboard. Use a water-resistant glue (such as a PVA medium) and let them dry thoroughly before using them.

Rinse the blocks clean after use, let them dry, and then you can put them away to use again another day.

Sometimes your child will want to print randomly, and sometimes he or she will want to try repeating patterns or printed pictures.

△ *Make your own printing blocks from sponge, cork or felt.*

Andy, aged three, used a circle and a triangle, the edge of a rubber and the end of a pencil to make this picture of fish eating weeds. ▽

Big Building

Children enjoy playing in tents and dens. Look carefully at real tents, sheds and houses, and make pretend homes in several different shapes.

Boxes

If you ever have a cardboard box from a big electrical appliance, like a washing machine, your child could use it as a house for a while. Let your child draw windows on the outside with a felt pen, then cut them out for him or her with a craft knife. Provide some cushions and a small blanket to make the house cosy inside, or make cardboard-box furniture. (See page 19 for ideas.) Decorate the outside of the house if you want to.

Which brick is the same shape as our tent?

There are many ways of making tents indoors. You probably have a favourite way of your own. Here are 3 suggestions.

Table Tent

Drape blankets, sheets or tablecloths right over a table, using safety pins (nappy pins are good) and clothes pegs to hold the corners together.

Drape a blanket over a table to make a tent.

Cooking

Make a 'campfire' or 'barbecue' by resting a cake cooling tray on two housebricks (or shoe boxes or tissue boxes). If you have any camping equipment yourself, you could lend your child a small billycan or camping kettle to play with, too.

Radiator Tent

If you have central heating, you may be able to use a radiator (when it is turned off!) to hold up a tent. Put one end of a sheet or a long tablecloth over the radiator, and hold it there with pushchair bag clips, clothes pegs, or by pushing two small cushions, one at each side of the sheet, partly down the back of the radiator.

Fix a sheet to a long radiator to make a narrow tent. ▽

Hold the other end of the sheet away from the wall by weighting it down with a row of big potatoes. This works best on a carpeted floor, as the sheet tends to slide about on a smooth floor.

Chair Tent

Tie some string between two heavy pieces of furniture, preferably so that it is about the same height from the floor all the way along. You could use two dining chairs, each weighted by a bag full of heavy groceries.

△ *You can make a washing line tent indoors by tying a piece of string between two heavy pieces of furniture.*

Drape a sheet over the string so that half the sheet is on each side, and fasten it with a few pegs. Hold the two ends of the sheet out by weighting them down with potatoes or tins of food.

5 More number games

In Chapter 3 we looked at how children learn to count and how they gradually become confident and familiar with bigger and bigger numbers.
I emphasised that learning to count takes *much longer* than many adults realise, and that practical experience of counting *real things* is extremely important.

In this chapter, too, the emphasis is on practical work with numbers. We shall look at a variety of ways in which you can help your child with learning about big and small numbers, and with the first steps in adding, taking away, multiplying and dividing (sometimes called the **'four rules' of arithmetic**).

Adding Up

How do children learn to add up? Like learning to count, it is a more complicated process than you might think, and different children go about it in different ways.

But here is what happened when I asked five different children the same adding question. Each child had an empty 'garden' and a saucer full of ladybirds in front of him or her (see pages 32 & 33) to work with, and I knew that they could all count to six accurately. Try this yourself with a few children if you can, and try to assess which stages they have reached.

How many ladybirds are there in my garden altogether?

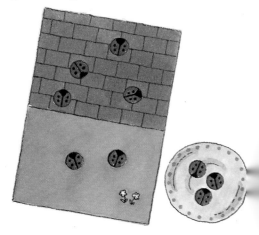

Children go about their adding in different ways as they grow in confidence.

Say to each child in turn:

'In my garden, there are 2 ladybirds on the wall and 4 ladybirds on the grass. How many ladybirds are there in my garden, altogether?'

Ben (aged 4) put 2 ladybirds on the wall, then 4 on the grass, then *counted them all,* from the beginning.
'1, 2, 3, 4, 5, **6**. 6 ladybirds.'

Jason (aged 5) put 2 on the wall, then 4 ladybirds on the grass, then *counted on from the first number* of ladybirds.
'**2**, 3, 4, 5, **6**. There are 6.'

Katie (aged 5) put 2 ladybirds on the wall and 4 on the grass, then she *counted on from the biggest number* of ladybirds.
'4 there: **4**, 5, **6**. 6 ladybirds.'

Sara (aged 6) did not bother to count any ladybirds. She *used her fingers* instead. She knows from her own experience that you can count 2 and 4 of anything, and it will come to the same number.
'**2**, then **4**, that makes **6** altogether. 6 ladybirds, then.'

Owen (aged 6) *gave the answer straight away.* He had counted 2 things and another 4 things so many times in the past that he knew the answer to my question would be 6, without doing any counting at all. If I had asked him a more difficult question, though, he would probably have preferred to use counting to answer it.

It is important not to rush children when they are learning to add. **Let them convince themselves of each number fact**, through repeated practical experience.

Adding Games

If you made the ladybirds, monsters in the bath, or snakes in the jungle (see pages 32 and 35) then you can use them for adding as well as counting.

Here is another game for extra adding (and taking away) practice.

Monsters in Dressing Gowns

You will need

* 4 pieces of card or stiff paper, about the size of this book.

* 40 miscellaneous buttons, which can be sorted easily into two groups (for example, 20 big buttons and 20 small, or 20 red ones and 20 of another colour).

* A pencil and felt pens or crayons.

* A box to keep them in.

Draw a monster in a dressing gown on each of the four cards. Colour each monster in a different colour.

Play '**Can you count . . .?**' and '**How many buttons?**' first (see page 34), to make sure you have a good idea of the largest number your child can count accurately and confidently.

Never try to do a sum where the answer is too big for your child to count to it happily and without confusion.

Keep to *very small numbers* to begin with, gradually introducing more numbers as your child makes progress.

How Many Altogether?

Start this game with a story like this: 'Monsters aren't fussy about whether their buttons all match. Blue Monster sewed some buttons onto his dressing gown. He sewed on two big buttons and one little button. How many buttons did he sew on, altogether?'

Count out the buttons, place them on Monster's dressing gown, and count the total. Then take it in turns to make up similar stories for each of your monsters.

Monsters aren't fussy about whether their buttons all match!

Give your child a variety of things to count.

Repetition and Variety

Repeat each sum with your child many times over a period of weeks. Try each sum with a variety of different things to count, too, so that she or he will realise that, for example, 'one add two makes three' is true *whatever* you are counting. Use fat snakes and thin snakes, ladybirds on the wall and on the grass, blue sponges and yellow sponges in monster's bath, green and red cars, and so on.

Number Facts and Number Bonds

'One add two makes three' is a **number fact** – it is a true statement using numbers. Some people call it a **number bond**, meaning exactly the same thing.

One particular group of number facts are those for multiplication. We usually call these number facts the **Times Tables**. Children need to learn these, as most parents will remember. But, of course, addition facts come first.

Lots to Learn

Just to give you a better idea of how much your child has to learn, here are some number facts for addition, written out like Times Tables.

(Do not be tempted to write down anything like this for your child to learn – it would *not* be helpful.)

$0 + 0 = 0$	$1 + 0 = 1$	$2 + 0 = 2$
$0 + 1 = 1$	$1 + 1 = 2$	$2 + 1 = 3$
$0 + 2 = 2$	$1 + 2 = 3$	$2 + 2 = 4$
$0 + 3 = 3$	$1 + 3 = 4$	$2 + 3 = 5$
$0 + 4 = 4$	$1 + 4 = 5$	$2 + 4 = 6$

Altogether, there are 121 different addition facts to learn, to cover every fact up to $10 + 10 = 20$.

There is no need to hurry, though. The number facts for adding should be learned gradually, a few at a time, and always in an enjoyable way using real things to count.

Number Families

A **'number family'** is a friendly name for all the number facts which make a particular number by adding.

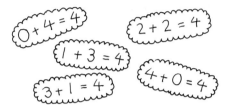

Here is the 'Family of Four'.

Practise each number family using any group of familiar and interesting objects. Here are some suggestions.

The Three Bears

Act out the story of a day in the life of the Three Bears when Goldilocks did *not* come to call. Find 3 teddies, 3 bowls and spoons, 3 cups, and 3 little blankets (teatowels will do). Tell the story, including plenty of counting.

'There were 3 bears asleep in bed. No-one was awake. Then Mummy Bear woke up, and got out of bed.

* So how many bears were still asleep?
* How many were awake?

There are three bears altogether!'

Play with the bears having a bowl of porridge and a cup of tea, going for a walk, and so on. Make sure you **mention all the combinations** of bears: no bears and 3 bears, 1 bear and 2 bears, 2 bears and 1 bear, and 3 bears and no bears.

Car Parking

Some toy garages have parking spaces marked on them, like the one in the picture below. You could make your own 'car park' as an alternative, just by drawing on a piece of cardboard. Use building bricks to make a wall around it.

Have just 4 cars to play with.

* How many cars are in the car park?
* How many are driving around?

There are four cars altogether!

Going for a Ride

Use five small toy people with a bus or lorry.

* How many people are there in the minibus?
* How many are waiting for a ride?

There are five people altogether.

How many people are in the minibus? How many cars are in the car park?

Six Skittles

Play with six wooden or plastic skittles, or use six plastic bottles, each with a little bit of sand in the bottom. Throw balls or bean bags at them to knock them down.

✱ How many did you knock down?

✱ How many skittles are there altogether?

Arrange the skittles in different patterns, to emphasise the numbers which make six.

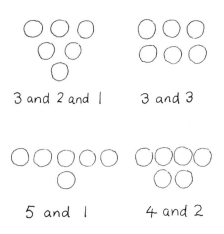

Seven Spiders

Make seven woolly spiders (see page 36) and draw or paint two spiders' webs in opposite corners of a big sheet of paper.

The seven spiders like to run backwards and forwards from one web to the other.

Throw dice to decide how many spiders will change webs.

Count how many spiders are on each web quite frequently.

✱ Can you get them all on the same web?

✱ Can you get exactly six spiders on one web, and one on the other?

✱ If there are four spiders on one web, how many are on the other one?

Make up other questions like these to ask each other.

Order Doesn't Matter

Your child will make a very important and useful discovery about adding up, after a while. She or he will notice that **it does not matter in which order you add two numbers, you will still get the same total**. For example, 4 + 3 comes to 7, so does 3 + 4.

'One day, 4 snakes slithered along into the jungle. Then 3 more snakes sneaked in. How many snakes were there altogether?'

'The next day, 3 snakes sneaked in first, then 4 more snakes came along. How many snakes were there altogether?'

How many snakes were there altogether?

Make up several stories like this to help convince your child that you can add up in any order.

Imagining

Gradually, as children become more and more confident with numbers, they will begin to be able to imagine numbers of objects, and to work out the answers to simple sums in their heads without needing the actual objects to look at and count. Obviously, this is a great step forward. Make sure you continue to encourage your child to use counting equipment, though, particularly as you tackle sums using bigger numbers.

The hungry teddy ate 2 apples.

Taking Away

Just as with adding, children need plenty of experience with real things to learn about **subtraction**. Use simple stories to provide a setting for each problem, in the same way as you have done for adding.

'Once there was a very hungry teddy. She had a bowl with 3 apples in it, and she ate 2 of them. How many apples were left?'

Use a teddy and some real apples to work out the answer by counting. Make up similar problems at each meal time, and use any of the counting and adding games you have made.

'Red Monster sewed 4 buttons on her dressing gown, but then 2 of them fell off. How many were left on?'

'Monster had 6 sponges to play with in the bath, but then he lost 3 of them. How many sponges did he have left?'

Families Again

A child who knows the 'family' of a particular number (see page 68) will find it easier to take away from that number. Make up some taking away stories to go with each number family activity you try. For example, 'There were 5 people who had been for a ride in the minibus. Two of them went off to have a cup of tea. How many were left behind?'

In effect, what you will be doing is showing your child that these sums are closely related to each other. Make up some stories to go with each number family.

$2 + 3 = 5$
$5 - 2 = 3$

$3 + 2 = 5$
$5 - 3 = 2$

Take Away One

Counting backwards is very useful when we take away just one, and many of us would also count backwards to take away two. (See page 40 for a few more ideas.)

Counting backwards is not usually the way most of us choose to work out a taking away problem. More often, **we count on** from the number we are taking away, **to see how many more we need** to reach the number we are taking away from. Many shop assistants do this when giving change. Alternatively, particularly when we are working with pencil and paper, we use the number facts we have learned.

Children are usually very quick at working out sums when something to eat is involved.

Varied Questions

Vary the way you ask questions.

* What's one less than 8?
* What is 8 take away one?
* What is one smaller than 8?
* If I've got 8 apples, and you have 7, how many more have I got than you?

Sometimes, try a mixture of adding and taking away questions. Let your child take his or her time to work out the answer, using any method he or she chooses. And, of course, make sure your questions are always set in context – in other words, that there is a 'story' to go with them.

Include some questions where the answer is *none*. For example, 'Here is Teddy's bowl of apples.' (Show your child a bowl with 4 apples in it.) 'Teddy will eat 4 of them for her dinner. How many will be left?'

Make up some puzzles like these where your child has to work out a 'hidden' number. Here are some examples.

Secret Sultana Eating

Collect together a bowl of sultanas (or pieces of carrot, peas or anything else you can count easily) and a plate. Count out a particular number of sultanas onto the plate.

'I've put 6 sultanas on my plate. Count them, to make sure there are 6. Close your eyes while I eat some . . . Now look! Can you work out how many I've already eaten?'

Hidey Bed

Choose some favourite teddies and dolls to play with. Count them carefully, and sit them in a row on the bedroom floor, next to the bed. Send your child out of the room for a minute, while some of the dolls and teddies hide themselves under the bedclothes! Can your child work out how many are hiding? Pull back the covers to check.

How many toys are hiding?

Ladybirds Under the Flower Pot

'There are 5 ladybirds in the garden. Count them, to check. Some of them are going to hide under the flowerpot. Close your eyes for a moment!'

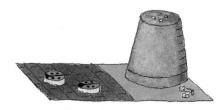

How many ladybirds are hiding?

Let your child lift the flower pot to check his or her answer. Take it in turns to choose how many ladybirds are in the garden, and to hide some.

Counting to 10

Ten is a very important number because our whole number system is based on 10.

Practise the 'Family of Ten' using your fingers. Start with 10 fingers up, and none down, and put your fingers down one at a time, counting 'How many fingers are up? How many fingers are down?' as you go. (Since it is difficult to put your little finger down on its own, start by putting your thumb down.) Then work in the opposite direction, putting one finger up at a time.

Play some 'detective' games starting with 10, too, like the games on page 72.

Pegboards

Use a ten-by-ten hole pegboard to show the ways you can make ten. Work one row at a time. On the first row, put in one peg in the left-hand hole. Ask your child how many holes are still empty in that row. Then fill in the rest of the row with pegs in a different colour. On the second row, put in 2 pegs of one colour. How many holes are empty? Fill in the rest of that row with another colour. Continue like this until the whole board is filled.

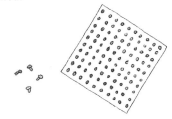

This ten-by-ten pegboard shows the ways you can make ten.

10 cars racing by Owen (aged 6)

Using Bigger Numbers

Once your child can count out ten, eleven or twelve objects accurately, she or he has obviously learned a great deal about counting. There is still quite a bit more to learn, though, and practical experience is still very important.

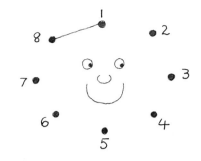

Practise bigger numbers with your children *saying* them, and *recognising* them when they are written down. (Later on, practise *writing* them – but not too soon.) Point out how carefully you have to listen, too, to avoid getting confused between, say, fourteen and forty.

Even with very small children, it is worth talking about large numbers, especially those which have a special personal interest. With numbers like '346' (our house number) make sure you sometimes say 'three hundred and forty six' instead of 'three-four-six', so that your child becomes familiar with *all* the number words we use. Look at house numbers, bus numbers, page numbers in books, and talk about people's ages.

Counting Never Stops

How will you know when your child's understanding of counting is reasonably complete? One of the important things your child will realise, probably some time after the age of six, is that you could carry on counting for the rest of your life and never stop. This is a sign that your child appreciates the pattern in our number system, and that he or she is well on the way to being a very competent counter.

Older children will like puzzles with more dots.

Younger children who recognise written numbers will enjoy joining the dots for very obvious pictures like this.

Dot-to-dot

Look for simple dot-to-dot puzzles for your child to try, or draw some of your own.

✷ Trace or draw a simple picture, then mark it with thick black dots.

✷ Cover it with a piece of thin paper, and hold the two sheets up against a window (in daylight).

✷ You should be able to see the dots through the paper quite clearly: mark over them with your pen.

✷ Number the dots carefully.

✷ Give the puzzle to your child to try. Make sure he or she joins the dots and not the numbers next to them.

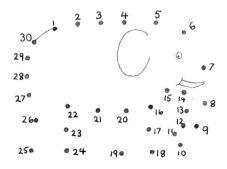

Adding Dice

Throw two dice and add the scores. You could do this with two small dice when you are playing a board game, so that you get round the board more quickly.

Seven Froggy Forfeits

If you have two large squashy foam or vinyl dice, you could play this game.

Each person has a turn at throwing the two dice. If the dots add up to 2, 3, 4, 5 or 6, the person who threw the dice has to do that number of froggy hops. If the total is seven, you *all* do seven froggy hops. If the total is 8, 9, 10, 11 or 12, then everyone *except* the person who threw the dice has to hop that many hops.

You can also play this game with croaking instead of hopping. Croak the same number of croaks as the total on the dice.

Velcro-tipped darts provide a safe game with plenty of adding practice.

Safety Darts

Use a safety dartboard with 3 velcro-tipped darts. This will give practice in adding numbers where the total may be more than ten. How could you score *exactly* ten with one, two, or three darts? Work this out for other scores.

Counting in Twos

Count in twos like this with a small group of children, using each child's name in turn.

2, 4, 6, 8,
Who do we appreciate?
RAPHAEL!

10, 12, 14, 16,
Say his name, don't keep him waiting!
RAPHAEL!

18, 20, 22,
Say it louder, all of you!
RAPHAEL!

Seven Froggy Forfeits is a good game for parties.

Counting in Tens

For each of the numbers up to nine, our number system uses a different symbol. Here are the symbols we use: 1, 2, 3, 4, 5, 6, 7, 8, 9. But we represent numbers bigger than nine by *re-using* those symbols, along with a 0 when needed. The *place* each symbol is put in tells us how much it is worth.

For example, in the number '257', the 2 actually stands for two hundred, the 5 for fifty, and the 7 just seven. But in the number '725', the seven is worth seven hundred, and in '572' it is worth seventy. Its value depends on the place it is in. (And this is what teachers are talking about when they mention **place value**.)

At first, children need to concentrate on numbers smaller than a hundred. Provide plenty of activities where you group things in tens.

Ten-bead Snakes

Use three strings and 30 beads. Ask your child to thread ten beads onto each string, then put them next to each other. Count the beads on each snake to make sure there are 10, then count all 30 beads, from 1 to 30, emphasising 10, 20 and 30 as you go. Make more snakes if your child wants to, and count further.

Count the beads on each snake.

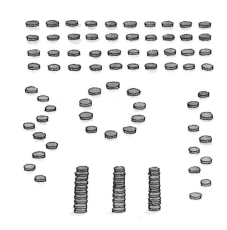

Most children enjoy playing with money.

Pennies

Collect a £1 bag of pennies from the bank, and tip them out on the table when you get home. (A plain-coloured tablecloth is a help.) Let your child play on his or her own to start with, perhaps counting, or making patterns, or piling the coins up.

After a while, ask him or her to help you put them in rows of ten, then count them all, emphasising the tens:
   1, 2, 3, 4, 5, 6, 7, 8, 9, **10**,
   11, 12, 13, 14, 15, 16, 17, 18, 19, **20** . . .
Count right up to a hundred. Then go back to the end of the first row, and count down like this:
   10, 20, 30, 40, 50, 60, 70, 80, 90, 100.

Next, push all the coins into a pile at the side. Let your child set them out in rows again, and count them again, both singly and in tens, as far as he or she can manage. Give plenty of praise and encouragement, and help him or her to finish if wanted.

More Tens

Point out to your child that counting in tens helps you to count bigger numbers of things without making mistakes.

✱ Bundle pencils into tens with elastic bands: put ten buttons in each pile on the table.

✱ Swap ten pennies for a ten-pence piece, and make amounts of money like 13p, 24p, and 40p, using both pennies and ten-pences.

Play a game where you can award yourselves ten points – for scoring a goal or knocking down a skittle. There is no need to compete with each other, as you can just try to beat your own score.

✱ Can you reach 50?

✱ How quickly can you reach 100?

Bean-Bag Clown

Make a Bean-Bag Clown from thick cardboard or a piece of hardboard about 60 by 75cm (2ft by 2ft 8in). Cut a hole as a mouth in the centre of the board (with a craft knife or jigsaw), a bit bigger than the size of this book. Paint a face on, using acrylic paints or emulsion or gloss paints.

Prop the clown against a wall or a chair. Throw bean-bags or rolled-up socks into his mouth, and agree to score a certain number of points for each one you get in. You can change the number of points you score to practise a different times-table. Score ten points for each bean bag if you are practising counting in tens, or two if you are practising counting in twos, and so on.

The Bean–Bag Clown provides good fun and practice at learning tables.

Multiplying and Dividing

Gradually, your child will build up a large repertoire of *number facts*. If these are based on his or her actual experience, and not just learned parrot-fashion, then he or she will be less likely to forget them and more likely to be able to use them sensibly.

Multiplication and division facts (they go hand-in-hand, like adding and taking away) are best learned by counting. Here are some ideas, using the counting and adding equipment.

'Here are the monsters, with no buttons on their dressing gowns. Then Red Monster sewed 2 buttons on her dressing gown. When her friend saw it, he wanted the same, so that's 2 lots of 2 buttons – makes 4 buttons altogether. Then the next monster wanted 2 as well. So that's 3 lots of 2 buttons, which is 6. Then – guess what! The last monster wanted 2 buttons as well. He didn't want to be left out. So that's 4 lots of 2 buttons, which is 8 buttons altogether. There were none, then 2, 4, 6, and 8.'

Work systematically like this, using equipment so that your child can count and check what you are saying. Use the word 'times' instead of 'lots of' some of the time. Make up stories about teddies who want cars to play with, or ladybirds flying into gardens, or snakes in the jungle. Practise each sequence of numbers with more than one sort of equipment, and work backwards sometimes – 8, 6, 4, 2, and none, for example, as each monster's buttons pop off, two at a time.

Do not try to tackle too much in one go. **Never use numbers which are too big for your child to count them easily**. It is best for your child to learn just a small part of each times table with confidence. For example, Sean aged 6, knows these sequences:

0, 2, 4, 6, 8, 10, 12, 14, 16, 18, 20.

0, 3, 6, 9, 12, 15.

0, 4, 8, 12.

0, 5, 10, 15, 20, 25, 30.

0, 6, 12.

0, 7, 14.

0, 8, 16.

0, 9, 18.

0, 10, 20, 30, 40, 50, 60, 70, 80, 90, 100.

Monster sewed 2 buttons on her dressing gown.

Puzzles

Make up puzzles for each other like these.

✳ 'There are four people in our family, and we all want 2 potatoes for our tea. How many potatoes will we need?'

✳ 'Each monster wants 6 buttons. I've got 18 buttons here. How many monsters is that enough for?'

✳ 'Here are 12 sponges for the monsters to play with in their baths. Can you share them out so that they each have the same number?'

✳ 'Here are 10 sponges for 3 monsters. Can you share them out fairly?'

Make sure that your child realises that not every number will share exactly.

Fractions

Real life often poses problems which involve sharing, and many can only be solved using parts of a whole thing.

'We've got 2 apples to share between four of us. How much can each of us have?'

Food is probably the best thing to use to learn about fractions! Cut up fruit, cakes, sandwiches and all sorts of other things into halves and quarters (and thirds, sometimes) so that your child becomes familiar with the idea of dividing something into equal parts.

Cutting up apples is a good way to learn about fractions.

Using a Calculator

When a child works out the answer to a problem like the adding and taking away ones we looked at earlier in this chapter, or like the 'Puzzles' on this page, he or she does not need to use the mathematical symbols $+$, $-$, $\times$, $\div$, or $=$, because no writing is needed. (Indeed, writing would often be a nuisance and a distraction, especially for under-6s.)

Using a calculator provides a very interesting way of introducing those symbols, though. If your child can count more than ten things accurately, and can recognise the numbers 0 to 9, then try a few simple sums together. Use just

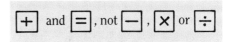

Show your child how to switch the calculator on and off, how to 'clear' ready for a new sum, and how the calculator writes each figure on its display.

Try sums on the calculator at the same time as ones with actual things, so that the calculator checks your counting and adding for you.

As your child gets older, encourage her or him to use a calculator. It is a valuable tool, not a threat to numeracy!

Three snakes and two snakes makes five.

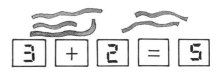

6 Measuring

We can measure all sorts of things, including temperature, length, weight, area, volume, value (in money), speed and time. Sometimes we only need rough measurements, perhaps to put things in order or to compare them, whilst at other times we need to be more accurate. Sometimes we measure things just out of interest, and at others we measure for a practical purpose.

Clothes provide an opportunity for using comparing words.

In this chapter I shall concentrate on **size** and **weight.** Many of the activities in Chapters 2 and 4 are also useful in helping children to learn about these.

Words We Use

Measuring is such a common part of our everyday life that we have dozens of words to use to compare things. Few of them have an exact meaning. For example, we found a slug in our garden which we all agreed was enormous; yet, of course, compared to us it was really very small. We were comparing it with other slugs we had seen in the past.

Here are a few of the most useful words:

big	small	wide	narrow
bigger	smaller	wider	narrower
biggest	smallest	widest	narrowest

and similarly with

| short | thin | full | high |
| tall | thick | empty | low |

| large | heavy | near | |
| tiny | light | far | |

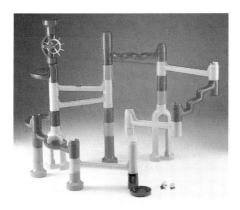

Can you make the helta-skelta track longer?

Marble Helta-skelta

∗ Can you make a long path for the marble to run down to the bottom?

∗ How can you make it longer or shorter?

∗ Can you make the helta-skelta taller or wider?

Clothes and Shoes

∗ Who has the biggest feet in our house?

∗ Who has the biggest shoes?

∗ Are my feet bigger than yours?

∗ Compare socks, shirts, pyjamas, coats, and gloves in various ways.

Wrapping Parcels

Wrap up various things in pretend parcels to give to each other.

∗ Is there enough paper?

∗ Do we need more sticky tape?

∗ Will the string go round?

Comparing Sizes

Talk about things you have seen and show with your hands or fingers how big or small they were.

∗ We saw some goldfish in the park and they were *this long*.

∗ There was a little beetle sitting on our step. It was only *this big*.

∗ We saw a teddy bear in the shop. It was bigger than mummy!

∗ Compare things with other things.

Encourage your child to make reasonable estimates of size.

Hop, Skip and Jump

∗ How far can you jump, or hop in one hop?

∗ How far can you throw a bean-bag?

∗ Measure the distance using your own feet by walking carefully heel to toe from start to finish.

∗ Try more than one jump, hop and throw. Which one was your best?

Bowls

Play a simple version of bowls. Place a marker to aim at (a smaller ball or a bean-bag or bucket). Have 2 or 3 balls each to bowl with. Try to roll each ball so that it is the nearest to the marker. You are allowed to knock other balls out of your way if you want to. Use a piece of string or wool to check which ball is closest at the end, if needed.

You can buy brightly coloured plastic boules sets in toyshops.

Far Away Things

Point out that when something is further away, it looks smaller. Watch cars and buses as they get closer, and compare familiar buildings when they are far away and when you are close to them. Emphasize that things *do not actually change size* when they are further away, they just look smaller. But don't expect your child to believe you at first!

Children need time and experience to establish for themselves that things (including people) do not change their size when they change position.

When my mum was on a lunchtime television programme for a few minutes, my daughter (then aged 4) watched excitedly. Next time we saw Nanny Rose in person, Tanya whispered to me, 'I'm glad they put her back the same size, then!'

Trick or True

One day, I was watching television with one of my sons, then aged 5. A magician was performing a trick using an ordinary pack of cards. Wherever he cut the cards the 8 of Clubs appeared. I could not imagine how he managed to do it – but my son was not impressed. He could not see why it was surprising that one card should keep re-appearing, partly because he did not know what the other 51 cards were. He had so little experience of using playing cards himself that he was not in a position to judge whether this was a trick, or just normal.

Sometimes children find it difficult to believe things which are true:

Which glass of orange juice would you like? There's the same amount in each.

Which piece of playdough would you like? There's the same amount in each.

Once again, what your child needs is more personal experience of measuring, using water, sand, and other things.

Water Play

Playing with water is very relaxing and enjoyable. Prepare well! Guard against people slipping on wet floors, have cloths handy for mopping-up when needed, and, particularly if you are using a bath or pool of water, organise yourself so that you do not need to leave your child alone.

Use different sorts of water on different occasions. Try warm water or cold water, bubbly water or water coloured with a few drops of food colouring (red, blue or green are good). Provide plastic spoons, cups, bowls, jugs and pots of various sizes, yoghurt pots, and stacking beakers and barrels. Transparent containers are especially useful (shampoo bottles, for example) as you can talk about them being empty, half-full or completely full.

Playing in the bath or outdoors in a paddling pool avoids problems with wet clothes because your child does not need to wear any.

Float cups in the water and pour water into them gently until they sink. Can you pour all the water from a little container into a big one? Can you pour all the water from a big container into a little one? Play with a water wheel. Can you make the wheel go round for longer with a big jug of water instead of a tiny cupful?

Playing on the draining board and in the sink, or at a table, allows children to notice water which they spill, rather than it just disappearing into the rest in the bath or pool. This means that they begin to be more accurate.

Find out how many teacups you can fill using jugs of different sizes, for example, or how many teaspoons of water you need to fill a little bowl. Pour water from tall thin beakers into short wide ones, to show what level it comes to!

Playing with water leads to many mathematical discoveries.

Sand, Buttons and Beans

It is sometimes useful to play with a large quantity of sand in a sandpit or on a beach, but also use sand in smaller quantities, to encourage your child to concentrate on comparing and measuring more carefully.

Silver sand is the best sort for playing with – it does not stain, and when it is dry it separates easily into fine grains. You can buy it at garden centres and some builder's merchants. (A bag of sand, as sold in a builder's merchants, is quite heavy. You may find it easier to buy an extra plastic sack, and divide it into two half-bags.)

Any clean paved area outdoors provides a good place to play with smaller amounts of sand. Put the sand in a washing-up bowl, and have a small dustpan and brush handy for your child to sweep the sand up when wanted.

Dry silver sand behaves almost like a liquid. You can tip it through a water wheel, and pour it into cups and containers of different sizes.

Dry sand can be poured like a liquid.

∗ Measure out tablespoons, dessertspoons and teaspoons full of dry sand into pots and dishes.

∗ Play with balancing scales. Put a cupful of sand in one pan, and watch the pan dip down because it is heavier than the other one. How can you make it balance again?

∗ Put some pebbles, bricks or cars in one pan, then fill the other with sand until it balances.

∗ Add cups of water to the dry sand a few at a time for your child to stir in and compare the wet and dry sand.

∗ Make mountains and mole hills, using big and small spades or spoons.

∗ Make some little sandcastles with yoghurt pots.

∗ If you use all the sand from three little sandcastles, will it fill your bucket to make a big sandcastle?

Many of the activities which you can do with dry sand can be repeated with *rice, lentils, dried peas* and *beans* or *small buttons*. Clean hands and clean pots and spoons to play with, will mean the foodstuffs are not wasted. Be vigilant, though! One of my sons once put a lentil in his ear, because he wanted to see if he could shake it right through his head and out of his other ear. The nurse in the hospital's emergency department said this is quite a common occurrence with four and five year olds, and it hurts.

Cooking involves plenty of measuring practice.

Cooking

Cooking and preparing food is a very satisfying way of learning about measuring. We use both more informal measures, like a *cupful* and a *level teaspoonful*, and standard measures like *grammes* and *ounces*. We measure time ('15 minutes in the oven') and *temperature* ('Very hot – 220°C or Gas Mark 8'). And then we eat the results!

Start with mixing ingredients in proportion.

* Dilute orange squash, showing your child how far up the glass to fill it with orange, and then fill with water.

* Mix concentrated fruit juice in a jug, with 1 part (tin, cup or mug) of juice to 6 parts of water, for example.

* Mix salads by the cupful, too. For example, try 3 cups of cooked rice, 1 cup of chopped celery, and half a cup of flaked almonds.

Read a recipe aloud, and show your child how you measure or weigh the ingredients. Very few children under eight will be able to read scales with a graduated dial, except possibly for measurements like 8oz and 250g, which are marked more clearly, but it is good for them to watch you. If your scales have weights, then your child may be able to help find the right ones and then balance them.

Try a recipe which needs grated cheese. Weigh the piece of cheese carefully, and show your child what the scales read or which weight it balances with. Weigh the cheese again after you have grated it, to show that it still weighs the same amount.

For any one recipe, keep to either *imperial* (ounces and pounds) or *metric* (grammes and kilos) measure, but help your child to become familiar with both.

Measuring is Useful

Older children will gradually come to appreciate the need for agreed standard measurements, particularly if they can watch you (and sometimes help you) measure for a real purpose.

Shopping

When shopping for food, point out things in different sizes. Shall we get a litre or a 2 litre bottle of lemon? Do we need a 1 pint or a 2 pint carton of milk? Shall we have 250g or 500g of margarine?

Talk about which things are heavy and which are light. Carrying tinned food and bags of potatoes is hard work, but loaves of bread and packets of cream crackers are easier to carry.

Which would you prefer to carry, the duck or the bucket?

Look how tall you've grown!

When shopping for children's clothes, talk about how they are labelled in sizes. Some shops use height. How tall are you in centimetres? Measure your child against a height chart or using a tape measure, to see, and let him or her help you measure another child, too. Some shops use ages for sizing. Which size are you, age 3 to 4, 5 to 6, or 7 to 8? Explain that clothes labelled by age are made so that they will fit *most* boys and girls of that age. Some shops use waist and chest sizes – are you 26″ or 28″? Let your child check the labels with you.

When shopping for shoes, explain that it is very important to get the right size, so that your child's growing feet do not get squashed. We have our feet measured, and we also try on the shoes to see if they are comfortable.

Decorating

Wallpapering, painting, fixing tiles, laying carpets and putting up shelves, all involve measuring. How many rolls of paper do we need? How many pints of water to mix this paste? How long should I cut this piece of paper? Do we need 2.5 litres or 5 litres of paint? Which width of carpet do we need?

As well as listening to you and watching you when you are measuring, your child can help you in a number of little ways. Let him or her hold the end of the tape measure while you read off the numbers. He or she can help you collect the right number of rolls of paper, and the tins of paint. And your child can see what is left over at the end. Real-life measuring often results in left-overs!

Travelling

Many young children think that the distance between two places changes, depending on which direction you are travelling in, and how fast you are going. Measure some shorter distances in steps, to begin to convince your child that the distance stays the same, and talk about any journeys you make, whether to the nearest bus stop or somewhere a hundred miles away.

Everyday Measuring

Many everyday adult activities rely on measuring. Learning to measure takes a long time – at least until the end of secondary school. Give your child as many chances as possible to measure things for a practical purpose at home.

7 Time

Time is a difficult concept for children, because they have comparatively little experience of time passing. It is very easy for we adults to forget that, for a young child, a year is a very big proportion of their whole life. Whilst for us, who remember 20 or 30 years passing, one year does not seem such a long time.

Shorter periods of time can appear to vary. An hour seems to pass very quickly when we are enjoying ourselves, but it can also feel interminably long if we are anxious about something or we are bored.

The seasons add further confusion. In the summer many parents have trouble persuading their children that it is bedtime because it is still light!

There are so many words for children to learn, too, and many of them are confusing. *Today* is not too bad, but *yesterday* and *tomorrow* seem to change their meaning every day! *In a minute*, *later on*, *morning*, *afternoon* and *evening*, *night* and *day*, *before* and *after* and *next week*, are just some of the words and phrases which need practising.

Let's look at how six children of different ages use words and ideas about time in their everyday lives.

Briony (aged 2) lives chiefly in the present. She only likes to wait a short time for something, but she understands sentences like 'You can have a drink *after* you've washed your hands'.

Andy (aged 3) uses words about time enthusiastically, but almost always wrongly! He says things like 'I'm going to playgroup yesterday', and 'Get up, mum, it's twenty o'clock'.

Emily (aged 4) knows how old she is now, and how old she was before that. She has a fairly clear idea of which order the regular events of the day take place in. She asks questions about the time quite often, like 'Is it time to get Daniel from school now?'

Simon (aged 5) knows that he does not go to school on Saturdays and Sundays. He knows whether it is morning or afternoon. He likes planning things. 'Can we go to the park tomorrow to see the ducks?'

Tina (aged 6) can tell you whether it is summer or winter. She knows where the hands on the clock will be when it is time to set off for school (8.30), even though she cannot tell the time yet.

Owen (aged 7) talks confidently about things in the past, present and future. He can tell the time if it is on the hour, half hour or quarter hour, but is still uncertain about other times. He knows that his birthday is on the 4th February, but not which year he was born.

Order of Events

The first thing that children need to grasp is that things happen in a particular order over a period of time.

Every Day

Talk about regular events in your child's life, and the order you usually do them in. Include getting up, getting washed and dressed, mealtimes, going shopping, visiting, and going to bed.

Help her or him understand *yesterday* and *tomorrow* by talking about things which *do not* happen every day. 'Did you have fun yesterday, when you went swimming?' or 'Let's phone up Maggy, and see if we can go and visit her tomorrow.' Talk about morning and afternoon, too. 'What did we do this morning?' 'What shall we do this afternoon?'

Jigsaws

Look for picture stories or jigsaws which need putting in order according to when things happened, like those shown below which were illustrated by Jan Ormerod.

Put these jigsaws together in the right order and watch events unfold.

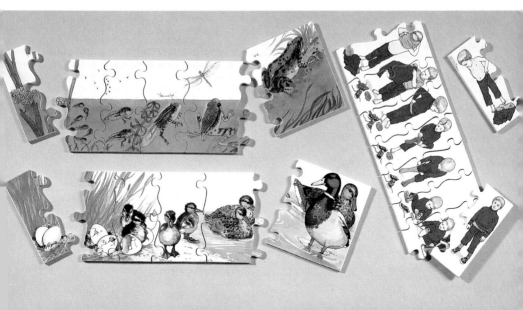

A Pretend Hospital

A pretend hospital gives you plenty of opportunities to talk about time. During one morning playing at hospitals, you can pretend it is morning, afternoon, evening and night-time several times.

Collect cardboard boxes as dolls' and teddies' beds, and large handkerchiefs and teatowels as bedclothes. The hospital needs a kitchen, too. (See pages 19 and 20 for ideas.) Paint or glue a picture and some controls on a small cardboard box, and tape a strip of card as a handle on top, to make a portable television. Choose some story books, a few jigsaws and some baby toys to keep the patients amused. Find cotton wool and a little dish for cleaning poorly arms and legs, and cut some short lengths of crepe bandage for them.

Your child can be both a doctor and a nurse. Find a white or light-coloured shirt to wear with the sleeves rolled up as a doctor's coat. Make a nurse's hat for him or her to wear from a strip of white card with a red cross drawn on it.

The patients' daily routine starts with waking up. Encourage your child to pretend to wash the patients' faces, and to give them breakfast. Doctor comes to see each doll or teddy after breakfast, then during the morning Nurse can read them stories, sing them songs and play games. Lunch next, then during the afternoon the patients can watch television or see visitors. After supper, it is time to go to sleep again.

It's time teddy was asleep.

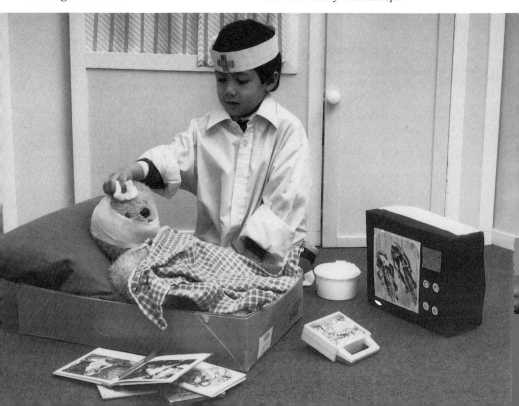

Photographs

If you have photos of your child at various ages from a baby to the present day, use them to talk about her or his past.

* That's you when you were a baby. You couldn't sit up yet, and you couldn't talk.

* Here you are with me when you were just two. You could run about, and talk, but you were still too little to wear pants – you had nappies.

* Here you are in the garden when it was snowy. Do you remember how cold it was? That was in the winter. It's summer now.

Give your child 3 or 4 photos to sort into order. 'Which one shows you as a baby? And which one shows you when you were two?'

Alarm Clocks

Explain what your alarm clock is for. Talk about what happens after the alarm has gone off, and what would happen if you did not wake up in time.

Wall Diary

Make a wall diary for one week. Use seven pieces of paper, and write the name of one day on each sheet. Draw a picture each day (you and your child between you) of something you have done on that day. It does not have to be anything complicated, just something your child will remember. Write a caption for each day if you want to.

For example:

Monday	We went to the post office. Jenny rode her tricycle.
Tuesday	We bought a melon.
Wednesday	Jenny went to playgroup.
Thursday	We went on a bus.
Friday	Jenny went to playgroup again.
Saturday	We had fish and chips.
Sunday	We went to visit Nanny and Grandad.

Fasten the diary to the wall as a frieze, and talk about it the week you are making it and during the following week. 'What happened yesterday?' 'What did we do on Monday?' and so on. Point to the relevant day's picture when you ask about it. Use the names of the days as often as possible, so that your child becomes familiar with them.

How Long a Time?

As well as learning that events happen in order, children need to learn about the way we measure lengths of time – in seconds, minutes, hours, days, and so on.

Eggs and Potatoes

Ask your child to help you time things which are cooking. Boil eggs and time them using a sand-timer, or watch the second hand on a clock or wrist-watch go round three complete turns. Bake potatoes in the oven for an hour, timed on a clock or with a kitchen-timer.

Use egg-timers and kitchen-timers to time other things. 'Let's set the timer for 5 minutes, and see if we can get all these toys cleared up before the buzzer goes.'

Familiarity with Time

Talking about the things you do in everyday life which involve time will help your child to become familiar with the words we use, and to see that time is important.

* Look for clocks while you are out shopping or walking. Show your child your watch, and tell him or her the time.

* Talk about important times. 'Come on, it's nearly ten to nine, you'll be late for nursery!'

* Listen to the radio or television. 'It's 9.30 a.m. on Friday 21st August, and here is the latest news.'

* Look up the times of television programmes in the newspaper, and find out how long they last. 'This programme is on for 15 minutes, so we'll have lunch when it finishes.'

* Look in your library books to see whether it is time to return them.

* Check the date, in your diary, on a calendar, or on a newspaper.

* Do some gardening together or grow things indoors to mark the passing of time and seasons.

* Talk about birthdays and other special dates.

Remember that your child needs to learn another group of number names (the *ordinal* numbers) for dates: first, second, third, fourth, and so on up to the thirty-first. 'We're going on holiday on the 19th of July. That's in two weeks' time.'

Growing things marks the passing of time.

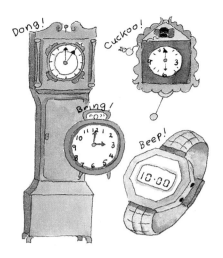

Can you tell me what the time is?

Calendars

Acquire a large calendar with an interesting picture for each month. Let your child turn over the page on the first day of each month, and talk about that month's picture, and all the ones before. 'What's the picture for February? It's a cat with her kittens!'

Having two calendars is doubly useful. You could have one with pictures related to the season of the year, and one with pictures of more general interest. Perhaps your child could choose a second calendar in February, when they are cheaper, to hang in her or his bedroom?

Telling the Time

Telling the time needs a great deal of knowledge. You must be able to recognise written numbers from 1 to 12, be able to remember which hand is the minute hand and which is the hour hand, and be able to count both forwards and backwards in fives, amongst other things. To add to the complications, some clocks have no numbers on them at all, and others (digital clocks) have no hands! Not surprisingly, telling the time is a skill which takes a long time to develop. It is best approached stage by stage, and should never be hurried.

Especially with under-sixes, **concentrate on actual times, not on ones you have set on a clock arbitrarily**. It can be very confusing for a child whose own concept of time passing is still developing, to sit with an adult who keeps moving the clock hands saying things like 'What time is it now? 4 o'clock. What time is it? 1 o'clock. Now let's make it 8 o'clock!'

Easy Times

Many five year olds can learn to tell 'afternoon times', which are the easiest times to learn: 1 o'clock, 2 o'clock, 3 o'clock, 4 o'clock and 5 o'clock. Have a clock in the room where your child is playing one afternoon, and look at it together every hour (if you remember!). A cuckoo clock or a chiming clock is helpful. Perhaps you could borrow one.

The best order to learn times is fairly obvious: o'clocks first, then half past, then quarter past, then quarter to each hour. Lastly, tackle minutes past the hour, then minutes to the hour. An old clock with hands which still work, or a geared teaching clock, is helpful.

You can have hours of fun helping your child learn maths through play. But remember that understanding is more important than parrot-like learning. Take your time and enjoy maths together . . .

Index

Adding up	64 to 70	Crayon Rubbing	38
Alarm Clocks	91	Cubes and Cuboids	53
Animal Counting	28	Cylinders	53
Baby's Body	23	Darts	75
Beads	48, 76	Decorating	87
Beakers and Barrels	18	Dice	75
Bean-Bag Clown	77	Dividing	78, 79
Big Building	62–63	Dot-to-dot	74
Birthdays	37		
Bowls	82	Egg Box Collection	27
Brio-Mec	57	Events in Order	89
Building Bricks	53		
Buttons and Beans	84	Family Languages	31
		Far Away Things	82
Café	20	Feeding Ducks	25
Calculators	79	Fish and Chips	24
Calendars	93	Fishing Game	30
Cardinal Numbers	26	Flexilink Cubes	57
Car Parking	68	Flip Book	40
Chalk Paths	43	Flip Fingers	18
Children Dancing	27	Flower Pots	72
Chime Ball	18	Foam Blocks	17
Circles	50	Football	43
Clothes and Shoes	81	Fractions	79
Conservation of Number	27		
Conservation of Size and Capacity	82	General Store	21
Conservation of Distance	87	Going Fishing	30
Construction Toys	53 to 58	Going for a Ride	68
Comparing Sizes	81		
Cooking	85	Hammer and Pegs	18
Count Down	40	Height Chart	86
Counting Collection	39	Hide and Seek	43
Counting in Rhythm	23	Hidey Bed	72
Counting in Twos	75	Home	19
Counting in Tens	76	Hop-scotch Trail	40
Counting Workbooks	25	Hop, Skip and Jump	81
Count on Your Fingers	29	Hospital	90

How Long a Time?	92
How Many People in Our Family?	27
Jigsaws	44, 45, 89
Ladybird Counters	32, 33
Ladybird Games	34, 64, 72, 78
Languages	31
Laying Tables	28
Lego	56
Linking Cubes	57
Make-Believe Play	19, 20, 21, 36, 62, 63, 90
Marble Helta-skelta	81
Me on a Puzzle	44
Mobilo	58
Monsters in Dressing Gowns	66, 78
Monsters in the Bath	35
Multilink Cubes	57
Multiplying	78, 79
Musical Statues	43
Number Bonds	67
Number Cards	39
Number Cut-outs	38
Number Facts	67
Number Families	68, 71
Number Frieze	40
Number Names	23
Number Spotting	37
Number Stories	24
Number Train	40
One to One Correspondence	25
One, Two, Three, Go!	24
Order Doesn't Matter	70
Ordinal Numbers	26, 92
Pass the Potatoes	28
Patterns	48, 49, 59
Pegboards	49, 73
Pennies	76
Pet Shop	36
Photographs	17, 44, 91
Playdough	20, 29

Playing Cards	39
Printing	59, 60, 61
Recognising Written Numbers	37 to 40
Rectangles	51
Reversals	41
Roads and Railways	58
Rubber Stamps	31
Safety Darts	75
Sand Play	47, 84
Scout's Pace	24
Secret Sultana Eating	72
Seven Spiders	69
Sewing	49
Shape Names	50
Shape-sorting Toys	51
Shopping	21, 86
Skittles	69
Snakes in the Jungle	35
Spot the Dog	29
Squares	51
Sultana Snacks	29
Taking Away	70, 71
Telephone Calls	39
Telling the Time	89, 93
Ten-bead Snakes	76
Tents	62, 63
Threading	48
Three Bears	68
Tissue Paper Numbers	38
Touching and Looking	47
Triangles	50
Triangular Prisms	53
Trick or True?	82
Varied Questions	72
Wall Diary	91
Water Play	83
Ways of Helping	46
What's in the Bag?	47
Whole Body Activities	43
Wrapping Parcels	81
Writing Numbers	41

Further reading

If you would like to read more about how children learn, these three books are especially recommended.

"Children's Minds" by Margaret Donaldson (published by Fontana) looks at research which shows that young children are capable of very complicated rational thought, as long as problems are presented to them in a human context, and considers some of the ways we can help children make the crucial transition to abstract thinking.

"Young Children Learning: Talking and Thinking at Home and at School" by Barbara Tizard and Martin Hughes (published by Fontana) describes a fascinating research project which compared conversations between four-year-olds at home with their mothers and with their teachers at school, and which showed that parents play an often-underestimated but very important part in their children's education.

"Children Learning Mathematics: a Teacher's Guide to Recent Research" by Linda Dickson, Margaret Brown and Olwen Gibson (published by Holt, Rinehart and Winston) is a comprehensive reference book which collects together information on how children aged 5 to 16 learn mathematics.